DATE DUE

GAYLORD			PRINTED IN U.S.A.

The NOVA Reader

The NOVA Reader

SCIENCE AT THE TURN OF THE MILLENNIUM

Edited by
Sandra Hackman

TV Books
New York

Publisher's Cataloging-in-Publication Data
The Nova Reader : science at the turn of the millennium / edited by Sandra Hackman. — 1st ed.
p. cm.
Includes bibliographical references.
ISBN: 1-57500-105-5

1. Discoveries in science. 2. Science—Forecasting.
3. Science—History. I. Hackman, Sandra. II. Title: Nova (Television program)

Q180.55.D57N68 1999 500
QBI98-1701

The publisher has made every effort to secure permission to reproduce copyrighted material and would like to apologize should there have been any errors or omissions.

TV Books, L.L.C.
Publishers serving the television industry.
1619 Broadway, Ninth Floor
New York, NY 10019
www.tvbooks.com

Principal Contributors

SANDRA HACKMAN was formerly acting editor and managing editor of *Technology Review*, MIT's national magazine, where she solicited and edited articles on the social effects of science and technology, including the environment, the economy, public health, and arms control. She is now an editor and writer based in Medford, Mass., where she explores the natural world with her husband.

BETH HORNING was formerly an associate editor at *Technology Review*, where she wrote about language and the mind. She has also published poetry and short stories and taught literature and writing at the university level. She is now a freelance writer and editor living with her family in Cambridge, Mass.

RICHARD MAURER is a science writer and children's book author—most recently of *The Wild Colorado*, an account of John Wesley Powell's second expedition through the canyons of the Colorado River, to be published early in 1999. His other books are *Rocket!*, *The NOVA Space Explorer's Guide*, *Junk in Space*, and *Airborne*, which won the American Institute of Physics Science Writing Award. He is writing the text for a Smithsonian traveling exhibit on the Hubble space telescope, and lives with his family in central Massachusetts.

PHILIP MORRISON is Institute Professor and professor of physics at MIT. A book reviewer for *Scientific American* since 1965, he is the author or co-author of over a dozen books, including *Nothing Is Too Wonderful to Be True* (1994), a collection of essays, and the recently released *Reason Enough to Hope* (MIT Press), which proposes redirecting nations' military budgets into a system of common security and development. He has appeared widely on radio and television, most visibly as narrator of the PBS series *The Ring of Truth*, which he created with his wife Phyllis Morrison, with whom he resides in Cambridge, Mass.

SUSAN REED is a senior editor at WGBH, Boston, and an award-winning freelance science journalist. She was formerly a senior science writer at Brigham and Women's Hospital and an editor and reporter for several Philadelphia newspapers. She lives with her son in Cambridge, Mass.

WILLIAM G. SCHELLER is an environmental, outdoors, and travel writer. A former editor of the New England Sierra Club's newspaper *Sierra* and the Appalachian Mountain Club's journal *Appalachia*, he has also contributed some 30 articles to *Sanctuary*, the magazine of the Massachusetts Audubon Society. Mr. Scheller is a co-author of *Adventures in Science*, NOVA's tenth-anniversary book, and collaborated with architect and photovoltaic design engineer Steven J. Strong on the 1987 book *The Solar Electric House*. He lives with his family in northern Vermont.

MICHAEL WALDHOLTZ is deputy editor for health and science at the *Wall Street Journal*. He led a group of reporters who won the 1997 Pulitzer Prize in national reporting for coverage of the new AIDS drugs. He is the co-author, with Jerry Bishop, of *Genome* (1990), and the author of *Curing Cancer* (1997). He lives with his family in New Jersey.

G. PASCAL ZACHARY is a science writer on technology, social issues, and international affairs for the *Wall Street Journal*, and also writes for other publications. He is the author of two books, *Showstopper! The Breakneck Race to Create Windows NT and the Next Generation at Microsoft* (1994), and *Endless Frontier: Vannevar Bush, Engineer of the American Century* (1997). He lives in London with his wife and two children.

Contents

Changing Our Lives

Paula Apsell

Executive Producer, NOVA

At the heart of science lies the process of discovery—uncovering what is hidden or unknown. The path to discovery requires curiosity, discipline, intellect, and the flexibility to respond creatively to the unexpected. In talking about discovery, physician-essayist Lewis Thomas said, "Surprise is what scientists live for. And the ability to capitalize on moments of surprise, plus the gift, amounting to something rather like good taste, of distinguishing an important surprise from a trivial one, are the marks of a good investigator. The very best ones revel in surprise, dance in the presence of astonishment."

For twenty-five years *NOVA*, the pioneering weekly series produced by WGBH-Boston for public television, has explored this process of scientific discovery. *NOVA* has gone beyond the headlines to reveal the men and women who engage in science and their far-flung fields of interest—from computers to genetics, from experimental archeology to deep space. This book offers you a look at some of the extraordinary advances in science and technology during the last quarter century, as well as speculation by top scientists and science journalists about what the future may hold.

A survey of all the scientific advances in the last twenty-five years would fill many volumes. We've chosen to focus on five areas that *NOVA* producers believe have had the greatest impact on people's daily lives: genetics, neuroscience, computers, the environment, and space. We hope that this book, like science itself, will leave you with more questions than answers, as well as an appreciation for the wonder of living in a world that is constantly being recast and redefined by new scientific insights.

Science Behind the Scenes

Philip Morrison

Institute Professor and Professor of Physics, MIT

Science and technology dance together across the entire stage of human history and prehistory, back to the first bang, the first fire, the first recognition of planets wandering the night sky. Both are part of human culture, neither solitary activities of lone individuals nor isolated islands of ideas.

Since World War I, when chemistry and the chemical industry came of age in the United States, American life has had more and more to do with the practice of science, though Yankee inventions and startups were surely important to world technology long before that.

After World War II, the swift rise of American science and its implied technologies were evident to all, reinforced by its mobilization, free from the havoc of war on our soil. A few magazines began to develop a culture of science for the postwar public; a number of the best newspapers followed with weekly science sections; and, after some years, *NOVA* came enduringly to public television. I reckon there have been several hundred hours of *NOVA*, which has likely given it the lion's share of input into bringing science to North American eyes and minds, not to mention the rest of the world. Science shows now regularly educate TV audiences by the millions; the other mass medium of image, film, has almost never used science as a theme, barring heroic biography and hyperbolic spectacles of disaster. (One can only hope that asteroid impacts are not as widely blamed on astronomers as the thermonuclear threat became the onus of physicists.)

The rise of medical science, and then of the computer in telecommunications and industry—especially in finance and commerce—have brought many tens of millions of Americans closer to technology and science, albeit in rather limited ways. The world is still run more by atoms than by bits, whatever the zealots say, but our discourse is certainly bit-borne. The computer on the desktop or in big mainframe racks has dwarfed all other links between science and the public, despite *NOVA*'s best efforts. But *NOVA* can and does show both the product and process of science, something that Windows XX and the Internet conceal well from all but insiders.

Today's software (maybe not the software of the day after tomorrow) punishes most approximations. Form is all: beware misspellings and case-sensitive entries. Of course these are human failings, more often stemming from the regiments of busy coders than from the user armies. Fortunately, the steady flow of new and usually better software makes clear what was wrong earlier.

That is only the most visible, if little discussed, instance of the great truth of science. It is not exact! Its successes are sooner or later transcended by better instruments, better data, better ideas. But the earlier ideas often survive as indispensable approximations. Is Newton absurd? Certainly not; the National Aeronautics and Space Administration today relies on Newton to navigate the solar system. Einstein's conception of gravity—a wonderful improvement on Newton's theory, and not final itself without quantum mechanics to explain the behavior of atoms—is complicated and therefore much less convenient than Newton's to handle the recurrent interactions of planet upon planet.

The science of light keeps ready a marvelous ladder of successive approximations, some two thousand years old, that are still taught and still useful. The ladder begins with the lore of straight light rays and shadows, of pinholes, eclipses, and their tricky combinations. The next step is the theory of light rays bent by mirrors, glass, air, and water, used for much optical design to this day, with computer assist. The science of light then progresses to wave interference patterns, indispensable for precision optical technology. These patterns were first modeled as to-and-fro waves like stretched Slinkys, then understood better as side-to-side waves like guitar strings. From there science moves to real electromagnetic waves, which give us radio, radar, and the rest. We next need to understand photocells and lasers in order to detect and transmit light, again using quantum physics to explain behavior on the atomic scale. Finally, the advanced particle physics of the 1980s unifies light, electrons, and the mysterious and nearly massless particles called neutrinos. Each of these theories includes the earlier approximations, although at real cost in the complexity of concepts and mathematics; all the theories, even the oldest—as old as Euclid—remain useful in their domains. Do you think the list will stop just now?

I strongly suspect one real surprise ahead, although I doubt that I will see it appear: the sciences of human society will strengthen greatly in the next generation. They will do so as their practitioners learn more, and as, indeed, those practitioners make the mathematical and experimental style of the natural sciences more their own, though certainly not by mere copying. The importance of language in human prehistory is already posing new and testable ideas, as is today's decisive slowing of the rate of human population growth. Most likely the next century will see all the sciences grow more than the last one did, but in paths that take new directions.

Let me leave this physicist's manifesto to move closer to the TV world. Take the present series of *NOVA* programs on experimental archeology, where you watch people you can easily recognize building obelisks and stadiums as best as they can with resources from the Public Broadcasting Service rather than from Caesar. You can see the questions raised vividly in stone and wood, not just as chalk drawings or computer printouts, and watch human hands and minds as they are set to solve problems. Participants work out step-by-step solutions to the degree required—not just what to do but why and how to do it—after watching what hasn't worked. No subtle instrumentation is needed; nothing more complex than squares, levers, ropes, rollers, oil, hammers...

The same approach can be used for avoiding war or grasping the actions of the cortex of the brain or forecasting the moves of the San Andreas Fault or projecting new cures for ills old and new. But solving those problems requires more steps, longer chains of logic, and often new instruments that can grasp and mold the tiny, the transient, and the hidden. All knowledge is partial and requires test; some answers are wonderfully useful, though even then they are generally incomplete and will be improved. This is the challenge for TV, and one *NOVA* has met: to show the process as well as the product.

Within the stream of Qs and As, we can expect that the right Qs and not some well-known As will be more important for moving on.

BUILDING BLOCKS

The Genetic Revolution: A Quarter Century of Biotechnology

Michael Waldholtz

Deputy Editor for Health and Science,
Wall Street Journal

In the mid-1970s scientists could only dream about finding the genes that are the driving force behind all Earthly life. By the early 1980s they were discovering a dozen genes or so a year. Today researchers are identifying dozens of genes each month, and during the coming decade they may churn out a similar number each week. The explosion of information from these findings is solving many of the deepest mysteries about how we humans carry out the moment-to-moment functions of our physical lives, and promises to shed light on how we think, how we create, and even why we dream.

By 2005, if not sooner, molecular biologists will have deciphered the complete chemical structure of the estimated fifty thousand human genes, as well as those that govern many microbes, plants, and animals. Painstaking exploration of these road maps will reveal many of the secrets of DNA, the single fabulous molecule at the center of every cell from which all genes are made. Indeed, commercial exploitation of the decoding of the human genome has already begun to yield more powerful and less dangerous treatments for devastating diseases.

One need look no further than the battle against AIDS to understand the potential impact of gene discoveries on human life. In the mid-1990s, fifteen years into the worldwide epidemic, few expected the imminent arrival of novel medicines to thwart the deadly action of HIV, the virus that causes AIDS. Yet in 1996, U.S. drug companies announced that combinations of powerful new drugs had restored life and vigor to thousands of HIV-infected people, many of whom had been on the verge of death. Since those drug "cocktails" have gone into widespread use, AIDS-related death rates have declined sharply, with some AIDS wards in hospitals shutting down and AIDS hospices cutting back their services in those countries where the expensive and cumbersome treatments are affordable.

Those drugs are the direct result of the 1986 discovery that one of the virus's genes is especially critical to HIV's ability to replicate itself and infect other cells. In the past, most drugs were found through the relatively crude trial-and-error technique of screening hundreds of thousands of chemicals for their medicinal value, often by testing them on animals with human-like illnesses. When one of those treatments turned out to work, scientists frequently didn't know why or how. But in the case of AIDS, by studying the protein called protease, which

W.H. Tobey/Courtesy of Harvard University Archives

Almost fifty years ago, James Watson helped discover the double-helix structure of DNA. In "Decoding the Book of Life," NOVA traced efforts to understand the molecule.

the HIV gene creates, scientists were able to design protease-inhibiting medicines to disable the newfound gene—a much more focused approach. While the first generation of protease inhibitors is probably the crude ancestor of medicines that may someday control the disease worldwide, the drugs' introduction and release ten years after the gene was found testifies to the speed with which powerful genetic therapies will continue to arise.

Similarly targeted gene-based discoveries have already yielded innovative ways to treat breast and colon cancer, prevent heart attacks, identify the causes of asthma and arthritis, and repair the defective genes that predispose people to common maladies. Perhaps no statement better describes the power of the new genetics than congressional testimony by Nobel laureate J. Michael Bishop in 1996. Dr. Bishop addressed himself specifically to cancer, but he could have been speaking about any of several other human scourges. "The human intellect," he said, "has finally laid hold of cancer with a grip that will eventually extract the deadly secrets of the disease. For the first time in my thirty years as a biomedical scientist I now believe that we will eventually cure cancer." Bold words, indeed.

But although they engender hope for longer, healthier lives, gene discoveries deserve to be treated with awe, respect, and care. Only a public educated about what is happening in biology laboratories across the United States and abroad can prevent potential abuses, many argue. For example, public officials are beginning to monitor the use of diagnostic tests that can detect genetic predisposition to diseases, and to pass state and federal laws to prevent insurers from denying coverage based on such tests.

In another cautionary note, the continuous and exuberant pronouncements by scientists and biotech companies claiming to have solved the mysteries of various diseases have turned out to prove simplistic and even wrong. In 1983, when scientists at the Massachusetts Institute of Technology revealed the structure of a gene that triggers uncontrolled cell division typical of a tumor, they brashly claimed they had found the key to cancer and were standing on the threshold of an entirely new way to attack it. Alas, the disease turned out to be a much more daunting foe: scientists were humbled as they found over the next fifteen years that dozens of cancer-related genes interact to produce tumors. Mapping all the biochemical pathways illuminated by these cancer-causing genes will probably take decades.

"The more we learn, the more we realize how extraordinarily complex the cancer cell is," says Robert Weinberg, a leading MIT biologist. "Each new understanding of the genes behind disease is in itself important and exciting. But understanding precisely what role the genes play and how best we can use that information to affect disease is going to be very difficult."

Each new genetic finding is also feeding the decades-old debate over the relative importance of nature and nurture, as biologists attempt to distinguish between which human attributes are inborn and largely unalterable and which can be affected by upbringing and environment. Many scientists argue that the complex interplay of genetics and environment may never be fully understood.

And the announcement by a Scottish scientist in early 1997 that he had created a biological copy of an adult sheep, in essence the first-ever mammalian clone, alerted the public to the fact that researchers were inventing biological tools that may be too powerful to comprehend or apply judiciously. Critics feared that scientists would soon unleash the ability to clone humans, laboratory-fabricated copies of living adults or even deceased individuals whose DNA had been preserved in test tubes. Some biologists responded that human cloning was a distant dream, and pointed out that a different environment was certain to produce a physically similar but unique person. But others maintained that human cloning was not only possible but might yield significant benefits—such as allowing infertile couples to conceive a biologically related child, or providing laboratory copies of organs for those in need of transplants. Some proponents even suggested that such scenarios might someday sound reasonable.

Human history, of course, is marked by scientific leaps for which society was ill-prepared, from Sir Isaac Newton's insight into the laws of gravity to Charles Darwin's discovery of evolution to the breakthroughs in physics that led to the splitting of the atom. Yet the record also shows that initially frightening techniques often become commonplace. Test-tube babies once seemed alien: critics warned that parents would select only embryos with a desired set of genetic traits. The ensuing search for perfect babies, these critics predicted, would prove corrosive to civilized life and perhaps even affect human evolution. Yet despite the alarms set off twenty years ago, schools today are studded with healthy, happy children created through in-vitro fertilization, with no discernible negative impact on modern life, at least not yet. In other words, today's scary biological scenario sometimes becomes a mundane part of the fabric of life in the not-so-distant future.

So it will be with the coming age of the gene.

Unraveling the Genetic Process

To grasp all this one must fathom a tiny chemical entity of almost unimaginable complexity and power. Most of what is known about the gene has arisen since the discovery of DNA, the critical molecule of which it is composed, in the 1940s. Francis Crick and James Watson then earned their fame by deciphering DNA's structure in 1953.

Genes lie along the several feet of DNA coiled tightly in an infinitesimally small package inside every cell. Each gene is composed of four subunits—so-called nucleotides, signified by the letters A, T, C, and G. A typical gene contains a long chain of perhaps thirty thousand letters in a specific order. Cells read this chain as if it were words on a page—a set of chemical instructions that direct the manufacture of protein molecules. Proteins are the key components in the engine of life, and their precise production is essential to human health. When a cell needs to conduct some bit of business—to reproduce or nourish itself—it sends signals into its nucleus calling forth specific proteins, which are then sent into action inside or outside the cell.

Insulin, for example, is a protein expressed by the thousands of four-letter combinations that make up the insulin gene. Other important substances that result when a cell reads a specific sequence of four nucleotide letters include the female hormone estrogen; the fatty protein cholesterol; and the enzyme TPA, which melts heart attack–causing blood clots. The body uses these proteins to spur growth of blood vessels, to ferry other proteins around the body, to make neurotransmitters that carry brain signals throughout the body, and to create thousands of receptors—tiny doorway-like substances that other proteins use to enter and exit cells.

Most biological functions that provide energy to cells, such as the body's processing of glucose, or blood sugar, entail numerous proteins interacting in a fixed chain of events. Should any of these proteins be altered or missing owing to a defective gene, the chain can be cut, and the normal process it rules crippled or completely shut down. For example, any interruption in the glucose-processing function can prevent the substance from entering cells easily. Barricaded outside the cells, the glucose backs up into the bloodstream and floods other organs and tissues. This glut can produce diabetes, a disease in which an individual has excessive sugar circulating in the bloodstream.

Why Are Genes Damaged?

Each time a cell divides to replenish itself or grow new tissue, it must make an exact replica of its full complement of three billion chemical letters. But, like humans, cells make mistakes, shuffling the nucleotide letters. This shuffling is actually an essential trick nature uses to differentiate one person and one species from another. Indeed, the altering of genetic structure through the copying process lies at the heart of evolution.

But sometimes mistakes during cell replication can create alterations so serious they cause troubling and even lethal health problems during an individual's lifetime. If serious mistakes occur in the cells that are female eggs or male sperm, they can create a so-called germ-line mutation, in which an altered version of a gene is transmitted to a newborn and sometimes even to succeeding generations. Genes can also be damaged by exposure to toxic agents in air, water, and food.

In an effort to investigate such effects, some twenty years after discovering DNA's structure scientists figured

Beach Media

This Seattle child, who lives in sterile isolation, is one of 20 million Americans suffering from a genetic disorder. Can such inherited maladies be cured? In "Genes by Design," NOVA probed current research and ethical implications of human genetic engineering.

out how to transfer a human gene into benign bacteria. These bacteria then manufacture the gene's protein while proliferating at a furious pace. Although our bodies normally produce proteins in tiny amounts, this recombinant-DNA technology has yielded vats of the body's treasure trove of natural substances for use against a wide range of ills. For example, doctors now use quantities of a hormone called human growth factor to treat undersized children, the human protein TPA to dissolve blood clots that cause heart attacks, and a protein called EPO to trigger the production of red and white blood cells in people whose disease-fighting ability has been diminished by cancer or chemotherapy.

More recently, scientists have developed a technique called polymerase chain reaction which, much like a high-speed Xerox machine, can quickly make tens of thousands of copies of DNA fragments. With so many copies available, superfast new computer-run machines then decipher the chains of nucleotide letters in the fragments. The effort to sequence the human genome's entire chain of three billion letters, which in 1995 scientists thought might take fifteen years, is moving so quickly that it may be finished by 2001. Scientists believe this detailed map will vastly speed up their ability to find genes and identify which alterations or variations make people more or less susceptible to disease.

Cracking the Code of Cancer

The impact of genetic alteration on proteins' ability to complete their prescribed jobs is the key to a fresh understanding of human health and novel techniques for attacking disease. Despite its complexity, cancer is one illness scientists are tackling most successfully with this knowledge.

Cancer, put simply, is an unwanted copying, or cloning, of a single cell. Until recently, the mechanism that causes a healthy cell to become a fast-dividing monster was simply beyond human knowledge. "When I started studying cancer in the early 1970s the cancer cell was considered a black box, a hidden mystery," says Bert Vogelstein, one of the world's leading gene hunters, whose laboratory at Johns Hopkins Medical School in Baltimore is credited with many discoveries regarding cancer genes. "But now we know that cancer is a genetic disease—a problem caused when certain genes inside a cell become altered or go awry. We are just now identifying these genes and the proteins they produce. Their discovery has become a beacon guiding us into a world that was impossible to investigate ten years ago."

Since the late 1980s, Dr. Vogelstein's lab, as well as others around the globe, has participated in a highly competitive race to identify the genes that play a critical role in transforming a healthy cell into cancer. These investigators will ultimately uncover perhaps several hundred genes whose job is to produce command and control agents—the protein molecules that tell cells when to divide and when to stop dividing. This doubling and tripling produces daughter cells that replace cells that die or slowly lose their functioning.

Scientists have found that cancer is caused mostly by defects in genes that produce the proteins that order a cell to stop dividing. In other words, cancer is simply the result of a missing or damaged protein that cannot carry out its task of turning off cell growth. Scientists have come to call these genes tumor suppressors. When the suppressor genes are altered, because they were either inherited in an aberrant form or damaged during life, cells normally under their watchful eye become renegades, massing bit by bit into a mob that can eventually become a deadly tumor.

"The understanding that cancer results when genes are damaged or mutated is the single most important step we have made in developing better ways to prevent, diagnose, and eventually treat the disease," says Richard Klausner, the director of the federal government's National Cancer Institute. "If there are cures in the coming decades, they will arise from this new knowledge."

The power of this model was first made explicit in a 1993 lab experiment involving a tiny slice of tissue removed from the body of Hubert H. Humphrey, deceased senator and presidential candidate. The saga actually began in 1967, when Senator Humphrey discovered spots of blood in his urine. His doctor sent off a urine specimen on a glass slide to several pathologists, who analyzed the shape of the cells to see if they were cancerous. Most decided the urine sample contained no evidence of trouble, but one, John Frost at Johns Hopkins Hospital, believed the cells he saw under the microscope revealed the very earliest stages of cancer. Having no way to prove his case, Dr. Frost was outvoted by the other pathologists.

Humphrey went on to lose the presidential race to Richard Nixon and returned to the Senate. A few years later, more blood in his urine led to exams showing that the politician did indeed have bladder cancer, and he died of the disease in 1977. But Dr. Frost retained the original urine sample, a dried bit of cells sitting on a glass slide locked away in a drawer, hoping he might someday prove he had seen the first signs of cancer ten years before.

In 1989 Dr. Vogelstein, in a different lab at John Hopkins, found that a particularly important gene, called p53, was often mutated in colon cancer tumors. He then discovered that p53 played a critical role in ensuring that cells do not divide excessively. He made this discovery by analyzing the genetic structure of cells in the colon as they passed through various stages, from a small growth called a polyp to full-blown cancer. Vogelstein's report that the critical change in the cell's genetics occurred when p53 was altered set off a worldwide rush to find p53 mutations in other cancer tumors. Scientists soon found that when p53 somehow became damaged, perhaps because of a replicating mistake or exposure to some environmental agent, the gene allowed cells in the breast, lung, brain, and bladder to grow out of control.

Dr. Frost, still trying to prove he had been right about Humphrey's cancer, turned in 1992 to a young doctor training in his lab who had worked with David Sidransky, a researcher who had helped Dr. Vogelstein find p53. Dr. Frost showed the slide to his student and asked him if it looked like the cells were cancerous. The student queried Dr. Sidransky. Several weeks later, the group of researchers obtained a tiny slice of Humphrey's tumor, which had been embalmed in wax. Extensive testing revealed that the slice of tissue contained cells with a p53 alteration. The researchers then returned to the urine sample slide and found that the p53 genes contained an alteration in the exact same spot as in the tumor.

"The detective work showed that indeed Frost was right," Dr. Sidransky says. "We had the evidence. Humphrey had the very early beginnings of a cancer when he ran for the presidency. Who knows how knowledge of that might have changed U.S. history?" As important, for the first time scientists could detect from a urine sample evidence of a tumor so small it otherwise could not be identified and caught. As with many cancers, bladder cancer can be cured if treated early. "If we knew in 1967 what we know now, Humphrey would never have succumbed to the disease," Dr. Sidransky says.

These days, Dr. Sidransky is developing a kit doctors can use to scan urine samples for p53 defects, a test that will certainly save thousands of lives. Indeed, cancer doctors are just now beginning to analyze the genetic makeup of tumors removed from many patients to determine what kind of genes have gone awry. P53 defects turn out to signify that a cancer is particularly fast growing, suggesting that physicians should attack it aggressively with surgery, chemotherapy, and radiation.

Preventing Deadly Diseases

This kind of predictive medicine is expected to become more widespread in treating and preventing diseases ranging from Alzheimer's to asthma to diabetes. Doctors will not only record patients' family history but also use a simple blood sample to study the makeup of their genes and determine what maladies they might contract, especially given a particular lifestyle or diet, or exposure to environmental agents.

For instance, scientists recently found that people who inherit a particular version of a protein that ferries fats about the body have a higher-than-average chance of developing Alzheimer's disease before age seventy. Those who inherit another variant of the same protein do not develop Alzheimer's until well into their nineties. Researchers aren't yet certain why these versions of the same gene can lead to different susceptibility to the tragic illness, but they hope that drug companies will develop medicines that can delay early onset of the disease.

"We think the best chance for treating Alzheimer's is slowing its development in people identified as carrying a high risk of getting the disease early," says Allen Roses, who discovered the protein variant. Working for Glaxo Wellcome, the giant British drug maker, Dr. Roses hopes to find a drug that can counter the proteins or chemical cascade triggered by the deleterious version of the gene.

The same search is occurring for cancer. When the Johns Hopkins researchers found that certain gene mu-

Courtesy Variety Artists

Michael Ochs Archives

Singer Arlo Guthrie (left) is a potential victim of Huntington's disease, the incurable hereditary illness that killed his father Woody Guthrie (right). NOVA reported on a test that can tell potential Huntington's sufferers their fate in "Confronting the Killer Gene."

tations lead to colon cancer, one doctor began experimenting with a drug called indomethacin, a painkiller used for treating arthritis that is closely related to aspirin and ibuprofen. Indomethacin works by interfering with the body's excessive production of inflammatory substances. Researchers had long believed that similar types of tissue swelling were involved in the growth of colon cancer cells. After preliminary tests in 1995 suggested that the drug may indeed retard or even reverse tumor growth, manufacturers began seeking even more potent versions. Thus, for the first time, the ability to identify people whose genes put them at high risk for cancer may enable their doctors to prevent the disease before it arises.

The discovery of two genes in 1994 and 1995 that cause an inherited form of breast cancer has similarly spurred a hunt for ways to head off the illness. "It is possible, we now think, to help women avoid cancers that might otherwise kill them," says Mary-Claire King, the University of Washington scientist whose twenty-year crusade, beginning in 1975, to unearth a breast cancer gene led to one of the discoveries.

Dr. King found the gene by studying families with unusually high rates of breast cancer. For decades doctors believed women in these families shared exposure to some cancer-causing agent in their air, water, or diet. "Few people, if anyone, thought cancer could be inherited," says Mark Skolnick, a gene researcher at the University of Utah whose lab found the first breast cancer gene, called BRCA1. "You have to hand it to Mary-Claire and the extraordinary work she did to track down the gene against some very high odds."

To definitively link a gene or set of genes to a disease, scientists must painstakingly identify families in which a gene-related disease appears to be present. After collecting these families, Dr. King and others used the high-speed DNA copying technique and computer-assisted sequencing machines to search for differences between the genes of women affected by disease and those of women who are disease-free. By developing pedigrees spanning three or four generations and retrieving DNA from as many as thirty to fifty family members, Drs. King and Skolnick pinpointed a tiny mutation in the genes of those with the disease. Research soon showed that 5 to 10 percent of all breast cancers stemmed from one of several mutant cancer-causing genes.

The technique spawned a rush to find families where illnesses such as cancer, heart disease, depression, and schizophrenia were present in much higher numbers than average. But skeptics questioned whether it was useful to test women for genes linked to breast cancer and other diseases if medicine could do nothing for them. Indeed, some women told that they carried a mutant gene decided their only recourse was to surgically remove their breasts. The breast-cancer gene created a "real quandary," says Arthur Caplan, a medical ethicist at the University of Pennsylvania. "Suddenly, we had to come up with new rules for dealing with very powerful pieces of information."

But scientists have recently stumbled across a finding that might sharply reduce the chances of breast cancer among these high-risk women. Doctors have long used a drug called tamoxifen, first discovered as a contraceptive, to treat breast cancer. The drug blocks the action of estrogen, the female hormone believed to promote the growth of certain breast and ovarian cancer cells. Some scientists wondered if the drug might actually prevent disease, too. In a highly publicized study, physicians in the United States and Europe recruited women who already had breast cancer, or who were at high risk because several close relatives had the disease or because they carried a gene variant that sharply increased their chances of developing the disease. These investigators recently reported a strikingly positive finding: women given tamoxifen had a 45 percent lower risk of developing breast cancer than those given a placebo, or dummy pill.

Within weeks of that report came news that Eli Lilly & Co. had serendipitously found a drug potentially even more potent than tamoxifen. Lilly and other companies had been working for years to develop drugs that could mimic or block the action of estrogen without the side effects of tamoxifen. Although tamoxifen not only blocks breast cancer but also prevents heart disease and even lowers the risk of bone-thinning osteoporosis among aging women, it also raises the risk of often-fatal uterine cancer. Lilly initially received approval for use of its drug, called Evista or raloxifene, to prevent osteoporosis but also tested it against breast cancer. (The drug mimics the bone-thickening characteristics of estrogen but inhibits the cancer-causing aspect of the hormone.) Only two weeks after the tamoxifen study was halted, Lilly scientists reported that Evista might cut the risk of cancer even more sharply than tamoxifen—without its side effects.

Calling Evista and other such drugs "the beginning of the era of designer hormones," Dr. Klausner of the National Cancer Institute maintains that "we may have the ability to consider and test drugs to lower the incidence of cancers the way we lower cholesterol." While fully exploiting the knowledge now at hand will keep scientists busy for decades, such excitement is understandable: for the first time they have a clear roadmap for studying diseases once considered inscrutable. Indeed, the twenty-first century will surely be known as the century of the gene because of the impact that unlocking its secrets, something regarded as unthinkable only a decade or two ago, will have on human life.

By identifying the genes that cause a cell to become a cancer, for example, scientists hope to identify the substances in the environment that can cause mutations in those genes. Armed with such information, policymakers could prevent—and individuals avoid—exposure to cancer-causing agents in water, air, and diet. Toward this end, a quick diagnostic test will reveal a newborn's particular genetic variants so parents will know whether the child is prone to obesity, heart disease, asthma, and even mental-health problems such as alcoholism and depression. Physicians practicing predictive medicine will then advise parents to eschew potentially dangerous diets or lifestyles, perhaps even prescribing an environment to deal with the child's specific susceptibility to disease.

In the not-so-distant future, scientists may actually repair and replace defective genes underlying disease. Indeed, some researchers are already substituting normal genes for patients' cancer-causing versions. However, determining precisely which cancers will respond to such treatments and whether they can be used without side effects will take years. And some scientists doubt that the approach will prove effective over the long term, since leaving untreated just one of a million cells from a tumor might allow it to return.

Gene-based inventions will include entirely new ways of fending off infectious microbes. For example, scientists at the Institute of Genome Research outside Washington, D.C., have recently mapped the genetic structure of the microbes underlying Lyme disease and syphilis, and of bacteria that can cause deadly infections in hospitalized patients. Suddenly "we have a blueprint for attacking these microbes in totally new ways," says director J. Craig Venter. Besides sequencing the human genome, Dr. Venter's goal is to unravel the genetic code of every major infectious bug, thereby giving drug makers fresh targets.

Some researchers forecast that in the coming gene age they may even be able to counter the biological causes of aging, allowing people to live longer and healthier lives. While all this may seem the stuff of science fiction, the exploding frontiers of genetic knowledge make it difficult to discount the seemingly impossible. Get ready, scientists say, for an unprecedented human adventure.

SCIENTIST'S NOTE:

Molecular Machinery

Robert A. Weinberg

Whitehead Institute
Daniel K. Ludwig and American Cancer Society Professor of Biology, MIT

Like many modern biologists, I work deep inside living cells, studying the molecules—DNA, RNA, and proteins—that make cells grow and divide, change their shape and form, and combine to form complex tissues. The ultimate secrets of life are hidden away inside these cells and molecules. Indeed, life's origins are still apparent in the molecules that our cells carry today.

My work is motivated by the assumption, now becoming more and more credible, that the forces driving malignant growth can be traced to how the molecules inside individual cells function in their normal incarnations to program cell division, and how they trigger cancer when they suffer damage. Cancer, like many human afflictions, is increasingly understood to be a molecular disease.

In the past, my colleagues and I have progressed very slowly in understanding how the mysterious growth-controlling molecules inside cells operate. These molecules talk to one another, forming complex signaling networks that resemble, at least in their outlines, the circuitry of signal-processing computer circuit boards. We biologists have fitted the molecular pieces of this circuitry together one by one. While doing so, we have uncovered the logic of how this circuitry organizes normal cells, and how it runs awry inside cancer cells.

The obstacles to elucidating the intricate wiring of cells are receding quickly owing to the advent of new biotechnologies. The Human Genome Project, DNA sequencing techniques, genetic strategies that teach us how proteins talk to one another, and the computerized ability to process and interpret this information will enable us to paint a detailed and highly accurate picture of how living cells function.

Many of the cutting-edge experimental techniques in use in 1998 were only vague dreams a decade earlier. The pace of technological innovation quickens; twenty-five years from now the biological techniques of 1998 will appear neolithic. Given this rate of rapid change, speculations on the biology of the distant future appear rather foolhardy.

But one long-term trend seems to have emerged already. Novel materials and instruments will increasingly be used to create virtually seamless interfaces between biological and electronic systems, interconnecting the molecules of life with silicon chips. Biologists will be able to analyze molecules, cells, and tissues in real time, and, ultimately, to manipulate biological entities by reprogramming the electronic circuit boards.

And so, in the course of time, we will reduce living cells to complex molecular machinery whose workings are understood in precise and intimate detail. With that reduction will come enormous power to correct the defects that lead to so many diseases, cancer being only one of them. The biological revolution has just begun.

SCIENTIST'S NOTE:

A New Perspective on the Origin of Life

Robert M. Hazen

Staff Scientist, Carnegie Institution of Washington
Clarence Robinson Professor of Earth Science, George Mason University

Twenty-five years ago the original source of energy for life was as certain as anything in biology. Every high-school textbook proclaimed what everyone accepted as intuitively obvious—that all life depends ultimately on the sun's radiant energy. Until recently, there has been little reason to doubt those claims. But new discoveries of life forms inhabiting ocean or rock depths far beyond the sun's influence have upset this comfortable certainty.

If there is one thing that science has taught us, it's that cherished notions about our place in the natural world often turn out to be dead wrong. We observe that the sun "rises" and "sets," but we now know that the Earth orbits the sun—we are not the center of the universe. Over many human life spans, mountains and oceans are unchanging attributes of our surroundings, yet we have learned that through the mechanisms of plate tectonics every topographic feature on Earth is transient over geological time: our war-contested political boundaries are doomed to disappear.

The great power of science as a way of knowing is that it leads us—haltingly but inevitably, through observations, experiments, and reasoning—to conclusions about the physical universe that are not self-evident. The history of science is littered with the overthrow of the obvious. Now, many scientists are betting that our intuitive view of life's original energy source is also in error.

Energy for Life

All living things require a continuous source of energy. Without energy, organisms cannot seek out and consume food, manufacture their cellular structures, or send nerve impulses from one place to another. They cannot grow, move, or reproduce. Until recently, almost all known life forms relied directly or indirectly on photosynthesis—the conversion by plants and a variety of one-celled organisms of the sun's light energy into the chemical energy of sugars, or carbohydrates. Carbohydrates can be used to build physical structures such as leaves, stems, and roots, or they can be further processed to provide a source of chemical energy for each cell's machinery.

While plants manufacture their own carbohydrates,

animals and other non-photosynthetic life must find another source of sugar. Thus we eat plants (or we eat animals that eat plants). There is an elegant symmetry to this story; the biological world seemed much simpler when the sun was life's only important energy source.

Our view of life on Earth was changed forever in the 1970s, however, when oceanographer Jack Corliss guided the submersible *Alvin* to the deep ocean's volcanic terrain of the East Pacific Rise. There, in the vicinity of "black smoker" vents that belch out hot, mineral-laden water into the cold ocean water, Corliss discovered astonishing ecosystems, with new species of crabs, clams, and bizarre six-foot-long tube worms. One-celled organisms also abounded, coating rock surfaces and clouding the water. These communities, forever cut off from the sun, thrive on geothermal energy supplied by the Earth's inner heat.

Microbes are the primary energy producers in these deep zones; they play the same ecological role as plants at the Earth's surface. These one-celled organisms exploit the fact that the cold ocean water, the hot volcanic water, and the sulfur-rich mineral surfaces over which these mixing fluids flow are not in chemical equilibrium. This situation is similar to the disequilibrium between a piece of coal and air—the coal combusts if it comes in contact with fire, releasing heat and carbon dioxide. Just as you can heat your house or power machinery by burning coal, so too can these deep microbes obtain energy by "burning" sulfide minerals or initiating any of a number of other energy-liberating reactions.

Following Corliss's revelations, dozens of other scientists are examining a wide variety of deep, wet environments. It seems that everywhere they look—in buried sediments, in oil wells, even in porous volcanic rocks more than a mile underground—microbes abound. These organisms seem to thrive on mineral surfaces where water-rock interactions provide the chemical energy for life. Such one-celled creatures account for only a tiny fraction of the rock mass, but the volume of Earth's wet crust is vast—billions of cubic kilometers. By some estimates deep life may account for half of Earth's total biomass. Our view of life has been skewed because these life forms are completely hidden from everyday view.

If so many organisms exist beyond the sun's radiant reach, then geothermal energy, and the chemically active mineral surfaces that are synthesized in geothermal domains, must be considered as a possible first power source for life. To be sure, sunlight remains the leading contender for life's original energy source, since the vast majority of known life forms do rely, directly or indirectly, on photosynthesis. In a series of groundbreaking experiments, University of Chicago graduate student Stanley Miller and his followers demonstrated that sunlight (as well as lightning and cometary impacts) can energize the conversion of simple gas molecules into carbon-based molecules, and thus provide the molecular building blocks of life. Thousands of studies have amplified these results, and a surface origin of life under a bright sun has become the seemingly unassailable conventional wisdom.

But nagging problems remain. Although most known species depend on the sun, solar energy can be harsh (as you know if you've ever gotten a bad sunburn). Sunlight can trigger the synthesis of smaller molecular building blocks of life, but its high energy tends to prevent the assembly or break apart the bonds of longer molecular chains, called polymers, on which all organisms depend. Furthermore, if the earliest life almost four billion years ago was confined to the surface, how did it escape the brutal, sterilizing bombardment by asteroids and comets?

The hypothesis of a deep, hydrothermal origin of life, far from asteroid barrages and harsh solar radiation, is supported by two recent types of experimental evidence. First, in studies of genetic mechanisms common to all life on Earth, University of Illinois biologist Carl Woese and others have revealed that microbes from extreme environments appear to be among the most primitive on Earth. Organisms that obtain energy from the interaction of water and rock may thus be closer to the first cell than photosynthetic microbes.

Experimental studies of organic synthesis in a hot, high-pressure water environment provide additional insights. Results from several laboratories suggest that vital organic molecules, including amino acids (the building blocks of proteins) and lipids (which form cell membranes), may synthesize in the presence of sulfide minerals. Larger molecular structures—the polymers essential to all known life—also seem to form more easily in the presence of mineral surfaces. The Earth's crust may thus provide an ideal chemical environment for the synthesis of life.

And there is another reason why we should look closely at the possibility of hydrothermal origins. If life is constrained to form in a sun-drenched pond or ocean, then Earth, and perhaps ancient Mars, are the only possible places where life could have begun in our solar system. If, however, life can originate in deep, wet zones, then life may be much more widespread. The possibility of deep origins raises the stakes in our exploration of other planets and moons.

The idea that life may have arisen in a deep, dark zone of volcanic heat and sulphurous minerals flies in the face of deeply ingrained religious metaphors. To many people, the sun represents the life-giving warmth of heaven, while sulphurous volcanoes are the closest terrestrial analog to hell. How could life have come from such a hostile environment?

Nature is not governed by our metaphors, however cherished they may be. Life as we know it demands carbon-based chemicals, a water-rich environment, and energy with which to assemble those ingredients into a self-replicating entity. We are many years from knowing how life began, but laboratory experiments under both surface and deep conditions, coupled with observations of life on Earth and, perhaps, elsewhere in the solar system, will be the ultimate arbiters of truth.

(This essay was adapted from an article in the *Planetary Report.*)

Scientist's Note:

Becoming Human

Ian Tattersall

Department of Anthropology,
American Museum of Natural History

The search for human origins holds unparalleled fascination for our egotistical species. Yet in the public mind, the thrill of this study by paleoanthropologists lies mainly if not exclusively in the uncovering of the fossil and archaeological records that alone reveal the facts of human prehistory.

Paleoanthropologists themselves have been as responsible as anyone for this view. Unlike scientists who study the evolutionary histories of rodents or whales or ants, categories that include a variety of species, paleoanthropologists focus on projecting a single species—*Homo sapiens*—as far back into time as possible. Yet spectacular recent additions to the human fossil record reveal an enormous diversity in the hominid family during the last four million–plus years. Instead of a single-minded slog from primitiveness to perfection, hominid history has evidently been a story of evolutionary experimentation, with new species constantly arising and exploring different ways to be hominid. Only because of some special and recently acquired qualities such as symbolic reasoning has *Homo sapiens* become the single hominid around. (For more on this process, see my recent book, *Becoming Human*.)

This new perspective means that clarifying humanity's origins and history will not entail simply discovering fossils but will also require deciphering complex events using techniques such as molecular genetics and computerized analysis of morphology.

My fervent hope is that during the next quarter century paleoanthropology will fall into line with the study of the evolution of all other organisms. Remarkable humans are, but we achieved our uniqueness through exactly the same mechanisms that gave rise to the rest of the planet's organic diversity. If we recognize this, we will reestablish paleoanthropology as an intellectual as well as a physical adventure.

BODY AND MIND

The Brain-Mind Connection:
A Quarter Century of Neuroscience

Beth Horning

Freelance Writer and Editor;
***Former Associate Editor,* Technology Review**

Back in the mid-1970s, when my friends and I wanted to get to the bottom of what might be bothering us, we would call up our most intimate memories from childhood and use them to explore the nuances of our upbringing and social conditioning. Today we do the same, but with a difference: sooner or later one of us is likely to say something like "I think part of the problem is just the way my head is wired."

It's a comment that bespeaks the growing influence of neuroscience, the study of the brain and central nervous system. Neurologist Richard Restak, author of *Brainscapes*, notes that over 90 percent of the neuroscientists who ever lived are living now, and that more has been learned about the brain in the past twenty years than in the previous two hundred.[1] Perhaps more significant than any specific discovery is the direction research is taking and its meaning for those of us embroiled in attempts to make sense of ourselves and our world.

Until recently, our knowledge of the brain came mainly from long, tedious examination of the effects of illness and injury upon it. The conclusions researchers drew were cautious, circumscribed, and of limited interest to the public, having to do with matters like which kinds of stroke damage diminish the ability to conjugate irregular verbs. Yet while such research continues today, a central question has come to the fore. Neuroscientists talk freely about how the brain might give rise to the mind—how the actual physical organ in our skulls could yield thoughts and feelings we think of as intrinsic to our personal experience. Far from eschewing psychological and philosophical issues, the field has come to regard them as a prime target of scientific investigation.

This new boldness stems partly from the technologies that have arrived on the scene, including positron emission tomography (PET) and functional magnetic resonance imaging (fMRI). These tools, which can provide something like a movie of brain activity, have yielded an array of provocative findings. For instance, fMRI was the force behind a 1995 media flurry about differences in the way men and women use their brain. Investigators at Yale University discovered that fMRI data from men who were engaged in a particular language task showed activity in a small region at the front of the left side of the brain. By contrast, the data from women engaged in the same task revealed activity not only in that area but in a comparable area on the other side of the brain as well. The *New York Times* quoted Sally E. Shaywitz, a principal author of the study, as saying that this could well be the first time "anyone has been able to demonstrate anything functionally different" between the brains of men and women.[2]

For their part, PET scans have been used to help undermine time-honored theories about the roots of obsessive-compulsive disorder (OCD), an affliction characterized by intrusive thoughts and repetitive, ritualized behavior. In its classic form, OCD may drive people to slavishly follow long, complex cleansing routines, plagued by irrepressible notions about how dirty they are. Psychoanalytical concepts have traditionally been invoked to account for such problems: informed that they have regressed to the "anal" stage of psychosexual development, patients have taken their place on the couch, there to hash out the barely remembered intricacies of their toilet training.

But the medical record has long hinted at an alternative explanation for the disorder. Since the beginning of the century, several organic brain diseases have been associated with OCD-like symptoms, and doctors have been able to relieve those symptoms by cutting nerve fibers that run between the brain's frontal cortex and an area called the basal ganglia, which lies beneath the two hemispheres. Interestingly enough, these are two of the main areas where today's PET studies of OCD sufferers show unusually high activity. According to Restak, most neuroscientists now think that OCD results from abnormalities in the nervous system's circuitry, plain and simple. In other words, the psychosexual development of people with OCD may be no more relevant than that of people who developed bothersome worries in the wake of encephalitis from the 1918 flu epidemic.

Exploring Consciousness

The sheer catalytic energy of certain scientists pursuing biological explanations for apparently psychological states—including even consciousness—has also propelled neuroscience forward. In an article for *Scientific American*, John Horgan has pointed to the key influence of Francis Crick, who shared the 1962 Nobel Prize for discovering the structure of DNA. In 1976 Crick moved to the Salk Institute for Biological Studies in San Diego, where he turned to neuroscience and eventually began to investigate the relationship between brain and mind. Horgan writes that "just as only the late Richard M. Nixon, famous for his red-baiting, could reestablish relations with communist China, so only Crick, who possesses a notoriously hard nose, could make consciousness a legitimate subject for science."[3]

And once the subject became legitimate, disciplines from nearly every corner of the academic universe established branch offices in neuroscience, giving rise to a new field of consciousness studies. Clearly, if members of

BBC

Outwardly normal, these men both have a puzzling brain injury that renders them unable to recognize faces. In "Stranger in the Mirror," NOVA *explored the nature of human perception.*

the scientific community once needed Crick's permission to probe the brain's depths, they certainly do not appear to need it now: a major distinguishing feature is the field's free-for-all atmosphere.

Crick has come to represent the "materialist" position, which is based on the belief that all the phenomena of the brain are governed by laws researchers can demonstrate through carefully designed experiments. He and his colleague Christof Koch, a neuroscientist at the California Institute of Technology, do much of their work at the cellular level, examining what happens when neurons, the cells that constitute the main working element of the brain, "fire," or transmit signals. "'You,' your joys and your sorrows, your memories and your ambitions, your sense of personal identity and free will, are in fact no more than the behavior of a vast assembly of nerve cells and their associated molecules," Crick writes in the introduction to his book *The Astonishing Hypothesis: The Scientific Search for the Soul.* "As Lewis Carroll's Alice might have phrased it: 'You're nothing but a pack of neurons.'"[4]

Crick and Koch's modus operandi has been to study visual awareness, since the system responsible for it is already fairly well understood. Brain scans have shown that many parts of the cerebral cortex come into play every time we see even one object. Crick and Koch have proposed that synchronized firing of the neurons in these different areas may yield a single unified perception. Such a process could explain why we are able to see a tree or a car or a flower instead of some odd collection of lines and color.

That question is the sort that can be addressed in the lab—by, for example, removing a piece of a living animal's skull, inserting electrodes into the brain tissue, and observing neuronal activity directly. Crick and Koch maintain that only when we understand how the brain works at this basic level can we take on grander issues; self-awareness and other riddles may well be soluble after we gain a more detailed nuts-and-bolts sense of just what neurons do. The two recommend that investigators leave more metaphysical speculations to "late-night conversations over beer" for now.

Philosopher and cognitive scientist David J. Chalmers of the University of California at Santa Cruz is not convinced by their basic argument. A materialist approach may someday fully explain the physical processes behind mental functions such as memory, attention, and yes, even self-awareness, he concedes, but it will do nothing to unravel the "hard problem" of why these functions are accompanied by subjective experience, why they elicit intensely personal feelings, thoughts, and associations. For instance, when we hear a song, we may remember a high school prom and experience a flood of disparate memories that we can communicate to someone else only piecemeal. The same song may prompt another person to think of a scene from a movie, someone else a long-ago day at the beach, still another person a remark a friend made last week over the phone, and not one of these people will be able to convey the exact quality of their experiences. Even something less evocative, such as a square of the color red, will elicit different associations in every human being, even though all may agree on what color they are looking at.

According to Chalmers, this reality constitutes the

true mystery of the mind. And he goes on to contend that once scientists grasp how profound the mystery is, they will be compelled to adopt a new way of thinking about consciousness. His view is that consciousness in the sense of subjective experience must be considered a "fundamental feature" of the universe; scientists need to treat it as an elemental concept like mass, time, or electrical charge, one of those entities that are central to our understanding of nature but are defined in only the most general terms.

Other students of consciousness take a wholly different approach to the study of mind and brain. One who has garnered much attention is Steven Pinker, director of the Center for Cognitive Neuroscience at MIT, whose book *How the Mind Works* is almost as ambitious as its title implies. Pinker, who clearly respects the materialists, also has a lot in common with philosophers like Chalmers in that he would vote for subjective experience as the mystery least likely to be resolved by neuroscience. But what really interests him is evolutionary psychology, the effort to wed cognitive science, which sees the mind as a system for processing information, with a Darwinian perspective.

Pinker argues that the mind is composed of "mental modules," each of which performs a specific information-processing function that controls speech or some other stupefyingly complex activity that average human beings perform with ease. Exactly what a given mental module does depends on what natural selection designed it to do back in the Stone Age, when our ancestors faced daily threats to survival. Pinker maintains we can determine that original purpose through something called reverse-engineering.

"In rummaging through an antique store, we may find a contraption that is inscrutable until we figure out what it was designed to do," he writes. "When we realize that it is an olive-pitter, we suddenly understand that the metal ring is designed to hold the olive, and the lever lowers an X-shaped blade through one end, pushing the pit out through the other end." Similarly, we might look at the multifarious talents and foibles common to our species with an eye toward how they could have helped us forage for food, outsmart animals, and in general adapt to a hostile environment. And when the purpose of those attributes becomes obvious, reverse-engineering can shed light on how the modules actually work.[5]

Take Crick and Koch's favorite, our sense of sight. Clearly an advantage for survival, it's an amazing ability if there ever was one. Pinker notes that every time we make heads or tails of what we're seeing, we're solving impossible problems in optics, as those who have striven to build sight into robots will attest. But if we consider what our sense of sight is for, we will begin to understand how we perform the feat: sight helps us make our way through life on this Earth. The process works because natural selection has designed our brains specifically to perceive things on a planet evenly lit by sunlight or moonlight and composed mostly of relatively smooth surfaces. No matter what we're looking at, our brain assumes the view is of that kind of world, so it has much less to figure out. When the lighting is uneven—that is, when we see shadows—we can be sure it is because some object is in the way, maybe a cloud, or a rock, or another creature. Impossible problems in optics become elementary.

In other words, we have a "surface-perception module" whose job it is to process visual information from the milieu where we evolved, and if we want further evidence that this module exists we need make only a slight change in our surroundings. "Say we place a person in a world that is not blanketed with sunshine but illuminated by a cunningly arranged patchwork of light," Pinker proposes. "If the surface-perception module assumes that illumination is even, it should be seduced into hallucinating objects that aren't there. Could that really happen? It happens every day. We call these hallucinations slide shows and movies and television."

The Era of Antidepressants

While students of consciousness craft animal experiments and attempt to explain subjective experience, other neuroscientists have demonstrated a link between brain and mind through contributions to modern pharmacology. Of special importance has been work on

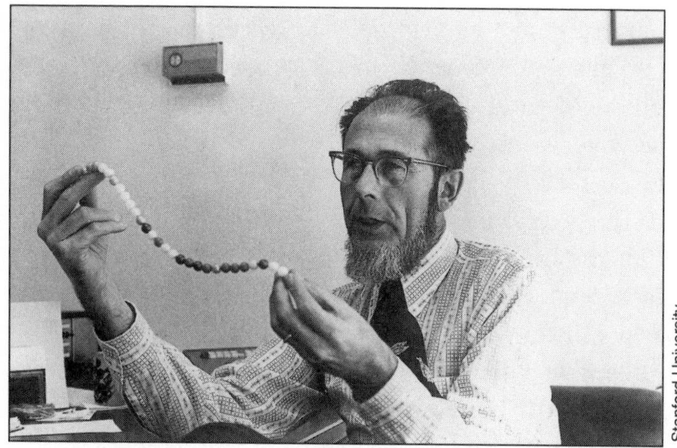

Stanford University

In "The Keys of Paradise," NOVA examined endorphins. Dr. Avram Goldstein, shown here with a model of an endorphin molecule, is among the scientists who argued that endorphins could revolutionize the treatment of pain, depression, and even schizophrenia.

chemicals known as neurotransmitters, which help convey signals across the synapse, or gap, that lies between neurons. Perhaps the greatest impact of this work has been in treating depression. Researchers have long noted problems with one particular neurotransmitter, serotonin, in the ailment. Throughout the 1960s and 1970s, and well into the 1980s, they pursued a drug for depression that would effectively target that chemical. In 1987 one finally hit the market under the name of Prozac.

The idea behind Prozac and spinoffs like Zoloft and Paxil is to keep neurons from sucking up too much serotonin, leaving more of it free to float around in the synapses and stimulate delicate receptors—the proteins on cells that receive chemical messages. Extra serotonin on the receptors usually relieves depression, although exactly how and why has yet to be revealed. One thing neuroscientists do know is that Prozac and the other "selective serotonin reuptake inhibitors" (SRRIs) do not disrupt the balance of other chemicals in the brain and so lack serious side effects, unlike the pills' predecessors. Some of the earlier antidepressants required patients to restrict their diets on pain of death—just eating a piece of cheese could cause blood pressure to skyrocket. As clinical experience with the drugs mounted and the list of prohibited foods grew longer, the idea of prescribing them for patients who, because of their illness, might not take good care of themselves began to seem downright dangerous.

Prozac's comparatively benign nature encouraged psychiatrists to put pen to prescription pad in cases where medication would otherwise have seemed a risk or a bother, and in less than three years the green-and-white capsule was on the cover of *Newsweek*. While some patients did not benefit much, a sizable percentage responded exceptionally well. Many began saying that it changed their lives. The antidepressant "seemed to give social confidence to the habitually timid, to make the sensitive brash, to lend the introvert the social skills of a salesman," Brown University psychiatrist Peter Kramer writes in *Listening to Prozac*.[6] He adds that patients who take the drug typically search their souls about the neu-

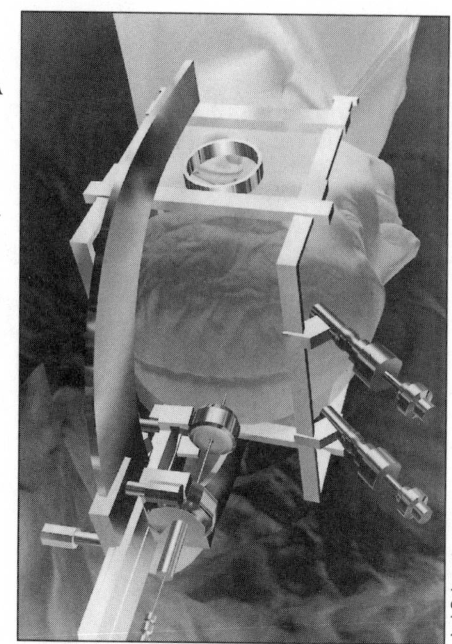

Using a computer graphic in "Brain Transplant," NOVA showed how fetal cells are inserted into the brain in a treatment largely unavailable in the U.S. owing to a ban on fetal tissue research.

Jed Schwartz

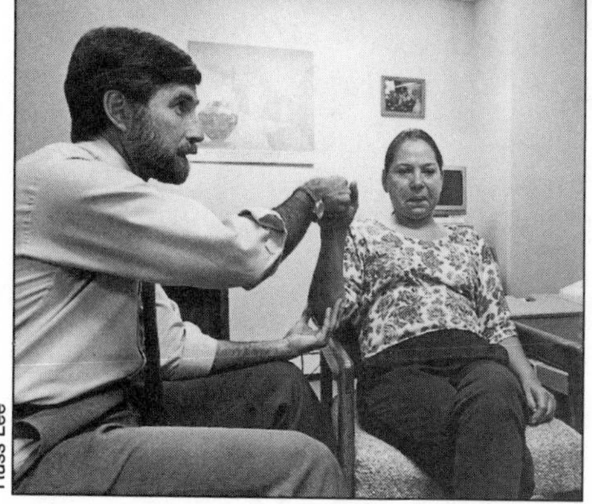

Dr. J. William Langston tests a woman who will travel to Sweden for a transplant of fetal brain cells to cure her paralysis in "Brain Transplant."

Russ Lee

rochemical transformation of the self: Prozac prompts them to ask which aspects of themselves are "real" and which are the result of bad chemistry they happen to have inhabited for much of their lives.

Consider the patient Kramer calls Tess. After her alcoholic father died and her mother descended into clinical depression, Tess, at age twelve, already a survivor of physical and sexual abuse, assumed responsibility for her nine younger siblings in the poorest public-housing project in the city. At seventeen, largely in an attempt to move her family out of the housing project, she married a man who soon proved to be an alcoholic like her father. Not surprisingly, Tess was miserable as an adult. Once her siblings grew up, she allowed her marriage to collapse and participated in a series of degrading affairs with married men. Yet professionally she was successful, having made a career of her talent for inspiring and nurturing others, and personally she was charming, evincing a caring manner and a ready sense of humor.

When Kramer put Tess on Prozac, at the time a brand-new drug, he did not hope to occasion a makeover. But he soon discovered that he had done exactly that. Not only did her deep sense of weariness abate as he had intended, but Tess was "no longer drawn to tragedy, nor did she feel heightened responsibility for the injured." Her romantic difficulties, almost inevitable in light of her past, vanished. Mentally, too, she was sharper, less prone to agonize over details, better able to make good decisions. Here was a woman who, whether on antidepressants or not, would seem to require a long, intensive effort to integrate traumatic life experiences into a coherent sense of self; in fact, that work was well under way. Kramer thought that what he was seeing suggested he would have to rethink his notions about just how long and intensive the process needed to be.

The drug's effects unsettled both him and his patient. Could so much of what had appeared to define her as a person really be such a straightforward matter of neurochemistry? "To find the Gordian knot dissolved by medication is a mixed pleasure: we want some internal responsibility for our lives, want to find meaning in our errors," Kramer writes. "Tess was happy, but she talked of a mild, persistent sense of wonder and dislocation."

When thousands of therapists see similar results in tens of thousands of patients, the entire approach to psychological problems begins to shift. Kramer notes that "drug responses provide hard-to-ignore evidence for certain beliefs—concerning the influence of biology on personality, intellectual performance, and social success—that heretofore we as a society have resisted." The result is that whether or not individuals who come to therapy leave with a prescription, mental-health professionals are likely to rely less on concepts such as the Oedipus complex to fathom patients' psyches.

Yet such a development prompts nagging questions. The Tesses of the world are not the only ones who could benefit from the confident, energetic, sociable way of being that SSRIs promise. In today's fast-paced culture, such traits are highly fashionable, and richly rewarded. Should we call any absence of them pathological and administer SSRIs as the cure? What kind of society endorses such a narrow spectrum of emotional hues as normal? Might SSRIs help people cope with stress only too well, so they never confront conditions they desperately need to change? "I have nothing against Prozac, but there needs to be a recognition that what may well be a rising tide of depression is related to the fact that basic social and psychological needs are not being met," says George Brown of the University of London, in a *New Yorker* article in which novelist Andrew Solomon chronicles his own struggles with depression.[7]

Granted, enthusiasm for SSRIs appears to have crested. Prozac prescriptions, which jumped by 35 percent in 1994, increased only 15 percent in 1995 and just 10 percent in 1996. New prescriptions for Zoloft have shown similarly slowed growth. Still, it's unlikely that any of this is occurring because people have a renewed desire to spend long hours sobbing in the company of mental-health providers. A more reasonable explanation is that they want to avoid even the relatively minor side effects of SSRIs, which can include weight gain, sleep disturbances, and loss of libido—not as bad as risk of stroke, but not such welcome changes, either. In the early days of SSRIs, when literature from pharmaceutical companies comprised more of the information available, those side effects were not widely recognized. Today the word is out, thanks to magazine articles and books like Cambridge psychiatrist William Appleton's *Prozac and the New Antidepressants*, not to mention countless conversations among friends and friends of friends on the drugs.

Significantly, side effects turn out to be something neuroscientific research is well equipped to handle. An infusion of funds stemming from the vaunted Decade of the Brain initiative, which the National Institute of Mental Health announced in 1990, has helped. Some of the most interesting results so far focus on the receptors, which have proved far more numerous than neuroscientists used to think. Serotonin, as it happens, locks onto some fifteen different receptors on neurons, not just one or two, and drug companies have every intention of exploiting that new knowledge to further reduce the side effects of antidepressants. More advanced medications will single out only those receptors that play a role in depression and neglect the ones important for regulating functions like sex drive.

"We're betting on this strategy," Edward Bradley of Sanofi, a leading health-care manufacturer based in Paris, told the *New York Times*.[8] Bradley, senior vice-president for clinical research, said that researchers have begun testing "medications specific for serotonin receptors in phase-one trials in Europe and will start testing them in the United States by the end of the year." If such trials are successful, the new breed of drugs should be available to U.S. patients by 2001.

Grappling with the Nature-Nurture Question

As we move into the next century, research in neuroscience shows no sign of slowing down. Psychologist Jerome Kagan of Harvard University believes that the field has entered a phase of discovery much like the one physics went through eighty or ninety years ago, when the components of the atom, once theoretical possibilities, were established as real. One result of all this investigation is often a kind of "biological determinism"—a view of the brain as a dictatorial entity affected by nothing except its genetic inheritance and the drugs born of our understanding of it.

That is improbable. Few psychiatrists would leap to the conclusion that because Prozac worked so well for Tess, her autobiography had nothing to do with her depression. Most are aware of researchers like biologist Robert Post, whose work with patients in the clinical wards of the National Institute of Mental Health has found that the brain may actually change anatomically in response to trauma. Some neurons may die, while others change shape; cells may make new connections or allow old ones to atrophy. The untoward events Tess and others have weathered could thus have rewired their nervous systems and made them more vulnerable to depression. If there's one thing that seems indisputable about the relationship between our experience and our physical being, it's that the two interact.

Then why do some scientists as well as members of the public appear to have such a hard time conceiving of it that way? Why does biological determinism stand a chance in the marketplace of ideas? Part of the answer might be that in earlier decades, the opposite proposition, what might be called cultural determinism, exerted undue influence. In his book *In Search of Human Nature*, Stanford University historian Carl Degler writes that for much of its past, American society, with its emphasis on bootstraps, has largely chosen to ignore biologically based characteristics, since they "might serve as an obstacle to an individual's self-realization."[9]

According to Kagan, researchers in neuroscience are mainly correcting the imbalances that have resulted from this way of thinking. His own pioneering work suggests that people truly are born with a biologically based temperament, a conclusion he draws from the myriad children he has studied. Although he himself has been careful to consider how his subjects' environment affects them and has stopped far short of biological determinism, he can see why neuroscience as a field has been more heedless. Like it or not, corrections are a common occurrence in science, he says, and "because all corrections go too far, neuroscientists have gone too far."

Anne Harrington, a historian of science at Harvard,

adds that the newly authorized question about how the brain gives rise to the mind is itself problematic because of its very power to fascinate. "The pull and the seduction of that question means that other kinds of questions are not being asked," she says. She points out that today's brain-imaging devices are a mixed blessing: neuroscientists gravitate toward research that these technologies are particularly good at addressing. The problem, she says, is that one can can be led to believe that the devices "provide an exclusive window into all relevant questions about our humanness."

Harrington further proposes that if political factors once made cultural determinism ascendant, they might also have something to do with today's drift toward biological determinism. "'Nature-nurture' debates" are "actually played out much more vociferously in the public arena than in science," she argues. The reason is that, depending on which side is ascendant, we make "certain choices in our own lives or certain judgments about our past choices that are not value-neutral."

For instance, if the nurture side gains credence, parents may blame themselves for their children's problems—feeling, for example, that their kids are poorly adjusted because of bad decisions on daycare. But if nature dominates the discussion, parents might feel obliged to accept the idea of Ritalin, which is often prescribed for children with neurological problems. In these sorts of situations, people want to think they have done the right thing. Unfortunately, when the atmosphere is so highly charged, "there's very little tolerance for complexity," Harrington observes.

But complexity is critical. Trouble arises from any description of the mind that excludes important perspectives, both she and Kagan maintain. That conviction guides them as co-directors of Harvard's Mind/Brain/Behavior Interfaculty Initiative (MBB), founded in 1992 partly to place advances in neuroscience in a social context. For example, MBB has convened a conference on the placebo effect, which Harrington says is "a wonderful kind of challenge" to the notion that our neurophysiology is immune to forces outside itself. "There do appear to be important biochemical changes in a brain that is exposed to a placebo," she points out, and what drives them is simply what people think is supposed to happen. To explore that phenomenon, MBB brought together not only neuroscientists and pharmacologists but also anthropologists and theologians to offer insight into the role that beliefs play in people's lives.

At the University of Chicago, biopsychologist Martha McClintock is similarly committed to addressing the questions that neuroscience seems to be neglecting. She will direct a new $12 million institute at the university, Allee Laboratories, that will investigate the ways people's behavior affects their biology and try to shed light on just how responsive to the environment the nervous system can be.

McClintock's own research shows that the messages the brain sends to the rest of the body can be determined by such elusive influences as smell: in one series of experiments, compounds swabbed from the underarms of young women at different times during their menstrual cycles have been found to either stretch out or compress the cycles of other women who sniff them. That such compounds exist for menstrual synchrony inevitably raises the question of whether they might exist for functions like, say, mating as well, either attracting potential sexual partners or staving them off. The results certainly indicate that studying the brain in isolation from the world has its pitfalls.

MBB and Allee Laboratories promise to pursue the kinds of complicated, subtle inquiries that do justice to the human brain. Kagan, for one, is hopeful about the outcome. The emphasis on biological determinism will not last forever, he says; neuroscience will eventually seek out a position somewhere between the extremes of nature and nurture.

How soon? Maybe in another twenty-five years. In the meantime, scientists and laypeople alike face a daunting challenge. Not only must they keep abreast of discoveries about the brain but they must also put those discoveries in perspective.

NOTES:

1. Richard Restak, *Brainscapes*, Hyperion, 1995.
2. Gina Kolata, "Men and Women Use Brain Differently," *New York Times*, February 16, 1995.
3. John Horgan, "Trends in Neuroscience: Can Science Explain Consciousness?" *Scientific American*, July 1994.
4. Francis Crick, *The Astonishing Hypothesis: The Scientific Search for the Soul*, Scribner's, 1994.
5. Steven Pinker, *How the Mind Works*, Norton, 1997.
6. Peter Kramer, *Listening to Prozac*, Penguin, 1993.
7. Andrew Solomon, "Anatomy of Melancholy," *The New Yorker*, January 12, 1998.
8. Daniel Goleman, "Next Generation of Psychiatric Drugs Will Be Quicker and Safer," *New York Times*, November 19, 1996.
9. Carl Degler, *In Search of Human Nature*, Oxford University Press, 1991.

SCIENTIST'S NOTE:

Mending the Injured Brain

Jam Ghajar

President, Aitken Neuroscience Center
Clinical Associate Professor of Neurosurgery,
Cornell University Medical College

Brain injury from accidents, usually involving motor vehicles, is the leading cause of death and disability among U.S. residents under the age of forty-five. This silent epidemic kills sixty thousand American citizens every year, as well as over one million people worldwide. And long-term care for the more than three hundred thousand Americans who survive the initial trauma, usually as bleak remnants of their former selves, costs over $40 billion per year.

The outlook for salvaging these devastated lives has long been dismal, as brain injury was once thought irreversible. But recent research showing that brain damage not only occurs at the moment of accident but also evolves over the ensuing hours and days has elicited hope. This research has revealed that most patients reach the emergency room alive but then die from swelling caused by fluid from blood vessels—a natural response to injury but devastating to a skull-encased organ. The injured brain is also exquisitely sensitive to a decrease in blood supply that a normal brain could easily tolerate.

Clinical research at the accident scene, in the emergency room, and in the intensive care unit has revealed that ensuring an adequate blood supply and oxygenation, and monitoring pressure in the brain, can significantly lower death rates while giving those who survive better-quality lives. But ensuring that this extensive research reaches all medical workers has proven to be a tall order: a national survey in 1991 by myself and other researchers showed that two-thirds of trauma centers were not monitoring brain pressure in patients with severe brain injury. The problem is that medical personnel suffer from information overload, and those in trauma care also labor long hours, often in the middle of the night, attempting to save lives on the brink of death.

To address this dissemination problem, ten neurosurgeons from around the country created *Guidelines for the Management of Severe Head Injury*, which summarizes the results of clinical trials and recommends the best practices. The Aitken Neuroscience Center in New York also developed an Internet database that allows medical personnel from the ambulance through rehabilitation to enter information on each patient. To demonstrate this system, philanthropist George Soros is funding a unique educational effort in Eastern Europe that will provide a template for the rest of the world. Neurosurgeons will use the resulting information, as well as the latest scientific findings, to continually update the guidelines and close the loop between research and clinical practice.

But even if such information sharing dramatically improves treatment for trauma patients, brain damage will inevitably occur in some cases. Because of this, as well as disease and aging in other patients, brain-cell loss will be the number-one cause of disability in the next century. The effects include the loss of self and social functioning in patients with Alzheimer's disease, and of motor skills in patients with multiple sclerosis, stroke, and Parkinson's disease. Although replacing brain cells could reverse these effects, the idea of brain transplants has long sounded like science fiction.

Fortunately again, that goal is becoming reality, thanks to work at a number of laboratories. For example, Dr. Steven Goldman of the Cornell University Medical College and the Aitken Neuroscience Center is now applying results from regenerating brain cells in birds to animals and humans. When removed from the lining of the deep cavities of the brain, so-called progenitor cells mature into several new lines of cells and make functional connections, given the right growth factors at the right time. The latest genetic sequencing technology will allow us to sort these cells into those that can potentially replace damaged areas in patients with multiple sclerosis, Parkinson's disease, stroke, and head injury.

The implications for better and less costly treatment are tremendous. Within the next ten years human transplants of functioning brain cells will become possible, providing a giant step toward realizing the dream of fully restoring the damaged brain.

Scientist's Note:

The Average Person

William Foege

Presidential Distinguished Professor of International Health, Emory University

My grandmother had ten children. Only five survived infancy. Five went on to become adults, one of those being my father.

When he was born, he was destined to die in 1953, given life expectancy at that time. But he saw all of the 1950s, all of the 1960s, and, incredibly, all of the 1970s. And then he saw all of the 1980s, and, amazingly, he has now seen the majority of the 1990s.

One of the most remarkable aspects of this century of science is not what happens in an emergency room or an intensive care unit or the laboratory. It's the information now available to the average person. The average person knows about tobacco and cholesterol, about the risks of the sun, and about using seat belts. The information given to the average person to run his or her average day can actually affect what happens to that person in the future.

At the beginning of the century we were struggling just to obtain scientific knowledge. At the end of the century we're struggling to respond to that knowledge.

NEW TOOLS

Human Ends, Human Scale:
A Quarter Century of Computing

G. Pascal Zachary

Senior Writer, Wall Street Journal

A quarter century ago computers were feared. Now they are seen as personal companions, suitable for children and essential to adult life. How did the shift occur? What transformed so profoundly the image and reality of the computer?

In the early 1970s, computers were large, impersonal machines that occupied entire rooms and required small armies of people to program and maintain them. While computers were no longer the special purview of the military, which had built the first digital computers after World War II and provided the bulk of research money well into the 1960s, they remained cloaked in mystery. Just posing a question to a computer took special training. Even then, an answer might not be returned for a day—or at all, if the questioner dropped one character from a long string of code.

Alien to ordinary life, the destructive potential of these machines obsessed the popular imagination. The image of the malevolent computer, growing impatient

Ed Hof/The Picture Cube

In "Computers, Spies, and Private Lives," NOVA *revealed how computers, routinely gathering information on our finances, politics, tastes, and habits, may pose a threat to individual privacy.*

with puny human intelligence, was indelibly imprinted on modern consciousness with the 1968 movie *2001: A Space Odyssey*. In the film, directed by Stanley Kubrick, the computer HAL takes control of an American spaceship and kills the crew.

Despite the outsized fears sparked by the computer, a more benign image of the machine sprouted in the shadows of this famously darker vision. An odd assortment of rebel engineers and counterculture sympathizers viewed the computer as a tool for thought, an aid to the overworked mind. They saw the computer not as an enslaver but as a freedom machine. Correctly designed and programmed, the computer would free humans from information overload, the bane of contemporary life, and from all sorts of political constraints.

The inspiration for this movement was a 1945 article written by Vannevar Bush, organizer of the Manhattan Project and the czar of military research during World War II. In the *Atlantic Monthly* article entitled "As We May Think," Bush described a futuristic desktop machine that could store a person's essential information and be flexible enough to record all the unique "associative trails," or links, that person made among pieces of information. While Bush's presumptions about technology proved wrong—he favored, for instance, reliance on microfilm and high-speed readers rather than electronic parts to store and retrieve data—his article provided a mental model, or map, that more digital-savvy engineers and inventors would follow.

In the 1960s a few engineers, notably J.C. Licklider and Douglas Engelbart, took Bush's vision as a starting point for their own work. Rather than consider the computer an impersonal number cruncher that did its work behind closed doors, they believed the computer's ultimate value lay in its capacity to display information visually. Not just text but graphics, sound, and even sensation were all within the reach of the computer. By combining the most compelling features of the movies and television with the sheer mathematical power of conventional computers, these engineers aimed to create a truly personal information appliance.

The barriers to realizing this vision, however, were immense. Computers were not only large but costly, and miniaturization took time. Beginning in the late 1950s, balky vacuum tubes had been replaced by tiny transistors, courtesy of a breakthrough by AT&T's Bell Labs. The transistor allowed computers to shrink in size and cost and grow in reliability. The next step was to shrink the transistor. In 1959 Texas Instruments debuted the integrated circuit, which permitted the miniaturization of many transistors on semiconductors, or silicon chips. Intel followed with a crucial innovation that obviated the need to wire together the transistors—the wiring was embedded in the silicon instead, saving time and shrinking the assembly even more.

The innovations in semiconductors made computers smaller and more powerful but still not small or cheap enough for personal use. This required another key innovation, the microprocessor, which crammed onto a single chip the basic elements of a complete computer. The microchip still needed to be connected to memory chips and input and output devices such as keyboards and display screens, but it otherwise acted as the traffic cop that regulated all the information zipping around the computer's circuitry.

Intel was the first to market a microprocessor, which it rightly dubbed a "computer on a chip." The chip's effect was monumental, giving hobbyists and garage engineers the means to build their own computers for the first time. In its January 1975 issue, *Popular Electronics* published an article describing a prototype for what is generally considered the first personal computer, a machine called the Altair that could be purchased in kit form for $400. The Altair packed as much power as other computers costing ten times as much. Its secret? It contained a microprocessor at its center, an 8080 chip from Intel. Incredibly, although Intel sold the chips alone for $360 each, MITS, the New Mexico company that made the Altair, obtained its 8080s for $75 apiece by buying in quantity.

With the Altair's arrival "the floodgates of personal computing opened up," in the words of historian Paul Ceruzzi. "The subsequent flood has washed over nearly every office worker and a sizeable fraction of homes in industrialized countries." While the Altair's achievement was at bottom technological–the company integrated some off-the-shelf parts—it also made a clever marketing move that foreshadowed the importance of hype in stirring consumer appetites for computers. In promoting its $400 computer, MITS carefully avoided saying that it wasn't much good without peripheral equipment such as a monitor. Though deceptive, the omission had the effect of attracting many novices to personal computing who might otherwise have been scared off by the Altair's intricacies. This was only the first time that a PC maker would downplay the true costs of adopting the machine. Understating the cost of the software, peripheral hardware, and training time required to master a PC was a tactic that peddlers of these dream machines would employ again and again in the years ahead.

The Software Revolution

Among the first to catch the PC fever was Paul Allen. A Seattle native, Allen was hanging around Cambridge, Mass., to be near his high-school pal Bill Gates, a Harvard undergrad. The two youths had already written software for big commercial computers, so when in December 1974 Allen found the article about the Altair in the latest *Popular Electronics*, he rushed to Gates's dorm room. On reading the article, the two focused on one compelling phrase: "The era of the computer in every home . . . has arrived."

These words intoxicated Gates and Allen, who saw an opportunity in the Altair: it came without software. Buyers had to write their own, or they were stuck with a useless computer. Gates and Allen decided to fix this problem by writing a language from which to create other programs. Before they even had such a language, they called MITS and promised to supply one. Then they took a language called Basic, designed by two Dartmouth College professors in the early 1960s, and adapted it to the Altair. Why they chose Basic (which stood for Beginners'

All-purpose Symbolic Instruction Code) was no mystery. The language was ideal for short programs and relatively easy to learn. And importantly, the creators of Basic asserted no ownership rights, so anyone could use or modify the language free of charge. Users could even turn around and sell their new variants of Basic.

Within six weeks Gates and Allen had written a version of Basic for the Altair and formed a partnership called Microsoft to peddle it. When the Altair became a hit, Microsoft's Basic sold so well—despite its $500 price tag–that Gates left Harvard. He never returned.

Other entrepreneurs would seize on the PC's potential, of course, but none gained the wealth, power, or notoriety of Gates. (His partner Allen also would become a billionaire but withdrew from managing the company before it became a household name.) As the industry grew, so did Microsoft. Its MS-DOS program became the chief software for IBM's original PC and the army of clones that followed in its wake. After a struggle with IBM over the direction of software, Microsoft again triumphed with its Windows program, which set a new standard for PC software partly because of its frank imitation of Apple's graphical Macintosh interface. The company even became the dominant supplier of word-processing, spreadsheet, and other applications for the Macintosh, the only credible rival to the IBM standard. Shrewdly playing both Apple and IBM against one another, Gates declared in

1984: "Macintosh is the only computer worth writing software for apart from the IBM PC."[1]

Gates's success lent credence to the idea that business factors had as much to do with the rapid spread of personal computing as did engineering and scientific ones. Standardization was key. This is the case with any new technological system—think of electricity, with the associated fight between alternating and direct current, or even the automobile, where in the early years of the century inventors of gas-powered engines battled backers of steam-powered cars. When a technology leaves too many choices to the consumer—should I choose beta or VHS for my videocassette recorder?—the result is often confusion, lower sales, less investment in the field, and a slower pace of innovation.

In this light, the drive by Microsoft and Intel to monopolize the basic elements of the PC, however devastating to rivals, proved a boon to consumers. Monopoly drove down costs through high volumes and turned the computer into a commodity. It made training easier and less expensive because at least the basic operations of a PC differed little from manufacturer to manufacturer. While computers exhibited more variety than, say, cars, the benefits of standardization were immense. By the early 1990s, these machines were no more exotic than VCRs and much more uniform than cellular phones.

By virtue of the PC's triumph, Microsoft is today the most powerful force in computing and one of the most profitable companies in the U.S., and Gates is the world's wealthiest individual (though his fortune is mainly on paper, consisting of shares of stock in Microsoft). Only Intel, still the dominant supplier of microprocessors, wields as much clout as Microsoft, as evidenced by *Time* magazine's choice of Andrew Grove, Intel's chairman, as its 1997 Man of the Year. Meanwhile, none of the fabled PC hardware entrepreneurs—from Apple cofounder Steve Jobs to Oracle chief Larry Ellison—have anywhere near Gates's wealth or renown.

Gates's status as the premier American tycoon underscores how central computing has become to the economy. But it also illustrates a signal shift in the field, which few predicted twenty-five years before. In the early 1970s, hardware was king. Think of the microprocessor and the hard-disk drive, which made computer storage affordable and ultra-reliable, as the twin pillars of the PC revolution. In those days, software was an afterthought. When IBM set out to market its original PC, it didn't even bother to produce its own basic software, or operating system, for the machine, choosing Microsoft as its supplier. That casual decision proved to be a monumental blunder because it robbed IBM of control of the basic look and feel of the PC. As hardware became increasingly standardized, software emerged as the only real way to distinguish a PC. IBM tried repeatedly to wrest control of software away from Microsoft, most famously with its OS/2 operating system. But consumers rejected OS/2, in part because it was designed to favor IBM machines, and over time Microsoft's dominance over software became entrenched.

Today, software is king and hardware is the handmaiden of code. Programmers, not the physical properties of semiconductors and electricity, define what is possible to achieve with computers. "This is a real switching of roles," says David Patterson, a computer scientist at the University of California at Berkeley and a leading chip designer. "Twenty years ago, hardware was brittle and software had to be simple in order fit within the limits of hardware. Now just the opposite is true. Hardware never breaks, and software crashes all the time."

The Internet Explosion

For all its wondrous qualities, the personal computer hit a wall in the early 1990s. The pause was probably inevitable. There was only so much people could do with a stand-alone appliance. Commercial innovation on desktop machines dried up, and universities failed to convert grand, blue-sky research into practical projects. The field was littered with grandiose dreams shattered on the rocks of reality. Artificial intelligence proved only that a computer could beat a person at chess but not recognize a baby's smile or drive a motor vehicle. Backers of natural-language translation, who once promised a machine that would automatically render Russian into English, gave up in disgust. Optical switches, which in theory would outperform silicon chips by relying on the speed of light to transmit data, remained a laboratory curiosity. Knowledgeable people began to fret that the go-go years of computing had ended.

As is often the case in the history of technology, the key to unlocking a new era in computers was lying around, relatively ignored. In the 1960s, the Department of Defense had quietly begun funding research into techniques for tying computers together through telephone lines. Part of the reason was practical: the brass wanted to ensure that information from all the sites could also be stored at each site, so data would survive if an enemy knocked out a machine. Such a network could also help in myriad other ways, such as by transmitting electronic messages among computers. The Pentagon's network was called Arpanet after its organizational sponsor (the Advanced Research Projects Agency). By 1971 the network connected twenty-three sites but was still basically invisible to the civilian computer community. To raise Arpanet's profile, the following year the agency held a public demonstration, which led to a surge in researcher interest. Still, the number of sites grew slowly. By the late 1980s, the Arpanet had mutated into the more civilian-sounding Internet (with the military creating its own separate network, Milnet, in 1982). Though popular with the computer cognoscenti, the Internet remained largely unknown to the wider public.

That changed rather dramatically with the advent of

the World Wide Web, which made publishing and distributing information on the Internet vastly easier. Unlike almost all the major innovations in personal computing since the 1970s, the Web was born in 1989 outside the U.S. Frustrated with the clunky, hard-to-use Internet, Tim Berners-Lee, a scientist at the CERN High-Energy Physics Laboratory in Geneva, combined two techniques for organizing information that together represented a leap beyond Internet convention. One technique was an addressing system, analogous to giving each phone line a distinct number, for finding files, pictures, audio, and video clips anywhere on the Internet. The other technique was a basic language, or tool, for combining such information into "home pages," and a set of codes for linking those pages together.

The Web turned the Internet into a mass attraction, launching a huge round of investment in companies that sought to bring publishing, retailing, entertainment, telephony, and a host of other activities into the realm of computing. While much of this entrepreneurial activity occurred in the U.S., the Web fired the imagination of the world. From a mere 313,000 computers in 1990, the Internet supported nearly 10 million by 1996. The success of the Web "holds a hidden lesson of its own," Michael Dertouzos, a computer scientist at MIT, has observed. "Technical proficiency means nothing to the general public. Ease of use and ease of posting their own information is what matters to users."

Such is the enthusiasm for the Internet that having a Web page of one's own has practically became a cliche. Not only do many businesses have one but some people do too. Pundits have declared a new era in human history. One booster, John Perry Barlow, has even gone so far as to describe the blossoming of the Internet as "the most important transforming technological event since the capture of fire." Political activists have begun referring to Web access as a basic human right, urging governments to ensure that the poor aren't left out of this shift in mass communication.

Exaggerated claims aside, the growth of the Web has clearly marked a new dynamic phase in the history of computing. But no one can say for sure whether the Web will become a fixture in human life in the manner of television and the telephone—or even whether the Web will remain as the dominant metaphor for the computer as network. That's because the Web has many shortcomings. It can be maddeningly slow. It is notoriously insecure, opening the way for credit-card scams and other fraudulent acts. And it seems to have only worsened the problem of information overload. Despite the Web, and indeed to some degree because of it, "we are drowning in information," has concluded *Interactions*, the journal of the Association for Computing Machinery, the leading computer-science group. The solution may be better tools for managing Web data, but these tools are years, if not decades, away. Today, says Andries van Dam, a computer scientist at Brown University, Web "retrieval systems are incredibly primitive."

Technical problems aren't the only ones burdening the Web. Commercial forces may prove even more formidable. So far, only a handful of companies earn a profit on their Web operations. To be sure, that's partly because they are still learning how to exploit the technology. But without more paid subscribers or advertisers, the Web pages of most media organizations will ultimately expire, and even retail sites won't survive unless consumers dramatically increase the volume of electronic commerce.

The Technology Bites Back

The hoopla over the Internet has enhanced the image of the computer as a feel-good technology that is blissfully carrying humanity into a digital future laden with information goodies and material prosperity. But this utopian vision tells only part of the story. Amid the din of relentless optimism, dissident voices have emerged. Computers have flaws and plenty of them, these critics claim.

Their howl has been reminiscent of an earlier era when big computers were seen as impersonal and carried the germ of tyranny. But the backlash against computers in the 1990s has been no reprise. The challenge is new. Critics can be grouped together around two big claims. The first is utilitarian: computers don't improve education and make work more efficient but actually hurt pro-

Betsy Bassett for WGBH

NOVA reenacted the story of how Clifford Stoll rooted out a spy ring infiltrating U.S. computers in "The KGB, the Computer, and Me."

ductivity. The second is spiritual or moral: computer-mediated experience dominates the contemporary scene and the effect, especially of the Internet, seems pernicious; by wedding people to simulated experience, the computer has impoverished daily life.

Economists and technology skeptics alike often point to the absence of statistical data showing productivity-enhancing effects of computers across the economy. Defenders counter that productivity gains take time to emerge, and also that these gains are simply missed by traditional measurements. Citing the scholarship of economic historian Paul A. David, they note that societies typically require decades to absorb new technologies, and that only after years of trying are people finally starting to use computers efficiently. Anecdotal evidence is growing—from the latest VCRs to speech-recognition software—that computers and the things they control are becoming easier to use and more effective.

Skeptics can point to disquieting counterexamples, however. Perhaps the most notorious is the so-called Year 2000 problem, which threatens to cripple computer systems critical to organizations from the Internal Revenue Service to garden-variety credit card vendors. To function properly, computers must keep track of time, so dates are critical. In the 1950s and 1960s computer memory was expensive, so to save space programmers set aside only two digits to represent years—1965 became 65 and so on. They never dreamed their programs would still be in use thirty and forty years later, but this "legacy" software proved durable and hard to replace. With the millennium approaching, the dating scheme must be fixed, or whole systems may shut down or simply refuse to recognize, say, a customer whose credit card expires in 2001. The problem is that a single company may have thousands upon thousands of date references imbedded in its many computer programs, so many, in fact, that some can't even be found. The cost of finding, and fixing these abbreviated dates (accomplished by adding "19" to the start of each one) is astronomical. Estimates range from the hundreds of millions to the trillions of dollars.

The Year 2000 problem is a classic "revenge factor": a seemingly trivial decision by programmers has "bitten back," to paraphrase historian Edward Tenner, who has studied scores of examples.[2] Such a process is also at work with the Internet, critics fear, suggesting that as more and more relationships are mediated by computers rather than occurring face to face, life loses some of its spiritual and moral value. In response to this specter, critics such as Clifford Stoll recommend a retreat from e-mail, the Net, and the accoutrements of virtual life.

But despite lingering misgivings about the computerization of everyday life, technological determinism alone fails to explain the role of computers in society. Their seemingly inexorable spread obscures evidence that people have more control over technologies than they think. Although the exotic physics of the microprocessor and the vagaries of software programs have greatly shaped the contour of computing, people's wants and needs, their habits and ideals, also shape machines and how people use them. The adoption of the telephone, for instance, provoked the same exuberant cries that social relations would never be the same, yet over time the technology bent under the weight of tradition and people's tastes. As sociologist Claude Fischer of the University of California at Berkeley has shown, the telephone chiefly reinforced existing social relations rather than disrupting them, thereby strengthening communities and helping people become more intimate with those they already knew.[3]

In the clamor over computers, the image of the all-powerful inventor, concocting gadgets in monklike seclusion from fellow men and women, reigns supreme. But the evidence is mounting that to a surprising degree people end up with the computers that they seek—not just what the engineers and scientists push at them. Think of all the "can't miss" computer technologies that never won a mass audience, including IBM's OS/2 operating system and "pen" computers that claimed to recognize individual handwriting but really didn't (few people wanted to scribble on their laptops anyway).

That some elegant technologies lose in the arena of public opinion is a reminder that the clout of scientists and engineers is limited. Consumers of technology push back, sometimes at least as hard as the inventors. This doesn't mean people get exactly what they want, or what they deserve, from a technology. But it does mean that we are squarely in an era when ordinary people will influence the future of computing—maybe even as much as any techno-elite. After all, computers are today so central to daily life that everyone has a stake in the way they work and the uses to which they are put. That stake may be expressed as a trivial concern about the way a mouse operates or a more philosophical stirring about differences in the way men and women use computers.

The larger point is that social, political, economic, and cultural forces invariably shape technologies in ways that give the lie to the notion that it is the machines themselves that drive changes in the lives of individuals and organizations. As journalist Robert Pool has observed, "Invention is no longer, as Ralph Waldo Emerson's aphorism had it, simply a matter of 'Build a better mousetrap and the world will beat a path to your door.' The world is already at your door, and it has a few things to say about that mousetrap."[4]

The Next Big Thing

Because of this, the only safe bet is that computer permutations will continue to surprise and confound people well into the next century, much as the Internet hit like a tidal wave in the 1990s. Yet some clues to the future direction of computing can be gleaned from the wizards in the laboratory, who still have a big say over the mousetrap. For engineers and scientists, much will depend on the rate of innovation in three critical areas of hardware: processing power, storage size, and speed of networking. "Moore's law" of computing—posited by Intel cofounder Gordon Moore—predicts a doubling of hardware speed and capacity every eighteen months. The stunning expansion that this law entails has already fueled the twin pillars of the computer revolution: declining prices and growing power.

Just how long Moore's law will hold up is an open question. The most vigorous debate has come in the chip field, the mother of the law. Some leading scientists think innovation is bound to slow dramatically over the next ten or twelve years if only because of the challenges presented by quantum physics, the unusual rules that describe the actions of matter at subatomic dimensions. By 2010, the tiniest aspects of a chip are expected to be smaller than 0.07 micron, or less than a thousandth the width of a human hair. At that dimension, electrons moving through chips begin behaving more randomly, busting through barriers placed by designers who realize that one rebellious electron can crash a switch.

Even Mr. Moore himself predicts large speed gains "for a couple of more generations of chips," which would take the field only into early next century. "Beyond that," he conceded in a 1996 interview, "things look difficult."

Still, chip designers have a million person-years of development and hundreds of billions of dollars of historical investment on their side—and a track record of defying naysayers. So for the foreseeable future, faster hardware at cheaper prices remains likely, meaning more dazzling computers than ever. In some cases what is now separate hardware, such as videoconferencing equipment, will be built into PCs. Screens won't just display vivid graphics but will convey the physical sensations of an experience. PCs will answer questions posed by users, not simply retrieve data that still requires time-consuming analysis. More-powerful wireless networks will enable people to do more of their computing on the move; through wireless modems, sunbathers will send e-mail and browse the Web at the beach. Computers in car dashboards (right above the tape deck) will ooze information about roadside attractions and store and display maps galore.

Computers will grow more powerful even as they become smaller, too, if scientists can draw on new materials and energy sources such as solar electricity and lighter, more efficient batteries. Mark Weiser, chief technologist at Xerox's research center in Palo Alto, Calif., has said, "The measure of the future of computing is whether you can hold a computer in your hand."[5]

Just what can be done with something so small? One idea: the credit-card computer. European banks already bury a microprocessor in charge cards; the chip has the power to store personal financial data and draw down electronic money, or "cybercash." The next step is to add a tiny radio antenna to the card, opening the possibility of uploading and downloading data from this bite-size computer to larger ones.

Researchers at Gemplus, a French pioneer of such smart cards, have a working prototype of a card-size computer with an ultra-thin display screen. Such miniature computers, bundled with radio sensors and an ultra-high-resolution screen the size of a quarter, might eventually cost less than $20, or more likely be given away for free by the provider of the service they access.

Cheap computers could become as common as buttons. The U.S. Army is experimenting with stitching a tiny computer into uniforms that acts as an electronic tag. This tag might be solar-powered and have the processing power of a computer that would have cost hundreds of thousands of dollars to build in the 1950s. Its purpose? To help sort laundry, by telling a clerk who the clothing belongs to. Such talking tags could also find their way into containers, boxes, and even Federal Express packages, easing the difficulty of tracking specific parcels stored in vehicles or warehouses.

Intriguing? Absolutely, but digital Nirvana is hardly around the corner. Despite Gemplus's prototype, J.P. Gloton, the company's chief researcher, admits that "today, this technology is really more of a dream." He says that scientists face big obstacles in creating lightweight batteries and both processor and memory chips that consume far less power than today. Nevertheless, if the past is prologue, engineers will likely break new ground.

Software is a dicier affair. Programmers always have found ways to squander processing speed and memory, no matter the advances. "We build ever faster and better generators, but the light bulbs and the electric typewriters come fitfully in their wake," geneticist Joshua Lederberg has metaphorically observed.[6]

Legions of computer scientists and hackers scratch their heads over the lack of software that fulfills the potential of the hardware. Some say code writers might even benefit from a slowdown in hardware innovation, since it would force them to exploit existing computers more intelligently. But under any circumstances, writing code is painstaking work that is still largely done by human hand.

Futurists talk bravely of computers someday producing their own software, gradually progressing almost organically to higher levels of sophistication. But to those laboring in the bowels of Microsoft or any other leading software maker, this idea seems absurdly optimistic. The big barriers to software breakthroughs aren't technical but practical. The costs of building a microprocessor are so large—a single factory can cost billions of dollars—that warring approaches are rare; standardization cuts costs and can increase innovation in peripheral parts. By contrast, software development remains relatively inexpensive because the equipment needed is so readily available. This opens the way for a confusing array of competing, and often incompatible, programs. Even worse, whereas people throw away their old hardware, thus freeing innovators to go pell-mell into the future, old software programs seem never to die. People have invested too much time getting accustomed to them, so each new generation of software must carry the whole history of the craft in its bosom. This makes new designs difficult.

But while inefficient software will squander some of the enormous computer power coming down the pike, there should be huge gains, too. Consider just one key field, speech recognition. The ability of a computer to meaningfully respond to human voices—and to talk back sensibly—seems on the verge of realization. "Speech is not just the future of Windows, but the future of computing itself," says Bill Gates, who predicts that speaking, not typing, may eventually become the chief way of controlling a computer.

Despite big gains in reliability, computers aren't quite ready to hold a conversation worthy of the movie *My Dinner with Andre*. The problem is immense. For a PC to take dictation, it must take into account different accents and dialects as well as people's tendency to employ metaphors, exaggerations, and humor. The fact that scientists barely understand how humans display such linguistic pyrotechnics makes the task of endowing a machine with such powers even harder.

But big money is being tossed at the problem. Besides

Patricia Hollander Gross/Stock Boston

This computer, programmed to play checkers, was able to defeat almost all its human opponents. NOVA examined the prospect of machines outpacing their creators in "The Mind Machines."

corporate research by the likes of Microsoft and IBM, the U.S., Japanese, and European governments are making massive investments in voice recognition. The effort seems to be paying off in a raft of experimental products. *Boston Globe* staffers "dial" coworkers by speaking their names into a telephone. United Airlines employees make their discounted seat reservations by talking with a computer program, which over time learns their preferences. With the aid of speech-recognition software, some fearless stock investors are buying and selling shares automatically, using the sound of their voice rather than tapping a telephone keypad.[7]

Whether it's speech recognition or computerized dashboards and credit cards, the frontier of information technology is sure to keep moving, as it migrates from the desktop to other unexpected places. Just as likely, the people served by these evolving machines will demand ever more of them. Can computers improve education? Can they ease the crushing burden of information overload? Can they really improve the productivity of labor—and make work more satisfying? More so than any purely scientific or technical issues, these are the questions that, for now, remain truly mysterious. Whether people end up fearing the computer again, as they did before the dawn of the PC a quarter century ago, depends on whether the wizards as well as the rest of us find a way to make this marvelous machine serve human ends on a human scale.

NOTES:

1. From *The Little Kingdom: The Private Story of Apple Computer* by Michael Moritz (1984), page 325.
2. Tenner explains this theory in *Why Things Bite Back: Technology and the Revenge of Unintended Consequences* (1996).
3. Fisher's analysis can be found in his classic study *America Calling: A Social History of the Telephone* (1992).
4. See *Beyond Engineering: How Society Shapes Technology* by Robert Pool (1997).

5. See *Prosperity: The Coming Twenty-Year Boom and What It Means to You*, by Bob Davis and David Wessell (1998), page 183.
6. For Lederberg's fuller thoughts and a broader discussion of the possibilities and limitations of software, see *The Future of Software*, edited by Derek Leebaert (1995).
7. See "Let's Talk: Speech Technology Is the Next Big Thing in Computing," *Business Week*, Feb. 23, 1998; "Gates Describes the Road Ahead as Noisy," *New York Times*, April 2, 1998.

FOR FURTHER READING:

On the military roots of computing, see *Creating the Computer: Government, Industry and High Technology* by Kenneth Flamm (1988).

For a study of computer rebels, see *Tools for Thought: The People and Ideas Behind the Next Computer Revolution* by Howard Rheingold (1985).

For the story of the transistor and the integrated circuit, see *The Chip: How Two Americans Invented the Microchip & Launched a Revolution* by T.R. Reid (1984).

The vast literature on Bill Gates is mostly fawning and occasionally irresponsible. The most reliable biography is *Gates* by Stephen Manes and Paul Andrews (1994).

For the definitive history of Arpanet, see *Where Wizards Stay Up Late: The Origins of the Internet* by Katie Hafner and Matthew Lyon (1996). In *What Will Be: How the New*

World of Information Will Change Our Lives (1997), Michael Dertouzos explains why the Web caught on so quickly.

For the case against computer-enhanced productivity, see *The Computer Revolution: An Economic Perspective* by Daniel Sichel (1997). For a good example of the argument that the computer is about to deliver a tidal wave of productivity gains, see *Prosperity* by Davis and Wessell (1998). For an interesting overall survey of this debate, see Jeff Madrick's essay in the *New York Review of Books*, March 26, 1998.

For a more thorough account of Stoll's argument for a retreat from the wired world, see *Silicon Snake Oil* (1995).

SCIENTIST'S NOTE:

A New Route to Artificial Intelligence

Patrick H. Winston

Artificial Intelligence Laboratory, MIT

From an engineering perspective, artificial intelligence is a grand success. Programs with roots in AI research perform feats of mathematical wizardry, act as genetic counselors, schedule gates at airports, and extract patterns from otherwise impenetrable piles of data. In those programs you find an armamentarium of AI tools hard at work, including tools for rule-based reasoning, exotic search, and neural-net training.

From a scientific perspective, however, not much has been accomplished: the goal of understanding intelligence from a computational point of view remains elusive. Computer reasoning programs still exhibit little or no common sense. Today's language programs translate simple sentences into database queries, but such software is derailed by idioms, metaphors, convoluted syntax, and ungrammatical expressions. The latest vision programs recognize engineered objects but frequently confuse faces, trees, and mountains.

Why so little progress? Since the field of AI was born in the 1960s, most of its practitioners have believed—or at least acted as if they have believed—that vision, language, and motor faculties are simply the input/output channels for human intelligence, and that a reasoning engine stands behind those faculties. Indeed, some adherents of this view suggest that people interested in studying vision, language, and motor faculties directly should attend their own conferences, lest AI gatherings be diluted by irrelevant distractions. Some of these practitioners even write textbooks that devote no space whatsoever to vision, language, and motor topics.

Of course, one could argue that thirty years is not much time for a science to develop. It might be that another thirty years, or another three hundred years, will be required to understand what lies behind vision, language, and motor functions. But there is a more attractive alternative: I believe our intelligence lies *in* our input/output channels, not behind them, and if we are to understand intelligence we must understand the contributions of vision, language, and motor faculties.

Evidence from Watching the Brain at Work

Researchers using functional magnetic resonance imaging (fMRI) and positron emission tomography (PET scans) can now determine which brain areas consume more energy as you think various sorts of thoughts. If you throw a ball, for example, your cerebellum lights up. If you watch the ball fly, your occipital lobe lights up. And if you hear the ball land, your temporal lobe lights up.

Nothing in those observations surprises: all are in accordance with classical theories of brain function. Remarkably, however, those same parts of your brain can light up without stimulation from outside events. For example, if you close your eyes and someone asks you to think of a letter of the alphabet, your occipital lobe lights up in a way reminiscent of what happens when you look at an alphabetic letter. Similarly, if someone asks you to think of a verb that goes with the noun *hammer* and you think of *pound*, several parts of your brain associated with understanding language light up, as does the right side of your cerebellum.

What? Your cerebellum? That part of your brain sitting on top of your spinal cord is supposed to be for fine motor control. What is it doing lighting up during what might seem to be a task for the language faculty alone? Such experiments force the conclusion that vision, language, and motor areas once viewed as used exclusively for processing external stimuli are involved in just plain thinking.

Ask a small child to add 2 and 2; she will convert the problem into a visual finger-counting exercise. Ask an adult to name the tenth letter in the alphabet; he will hold up his hand and start counting, just as if he were a

AP/Wide World Photos

World chess champion Gary Kasparov shown beating a computer at chess in NOVA's "The Chip and the Chess Master."

child adding. Ask a physics student to solve a problem; she will draw a diagram. There is no doubt about it: vision makes it possible to solve problems that would otherwise be difficult or impossible. To be sure, visual problem solving is not the only kind, but it is hard to imagine how we can understand how the brain thinks unless we understand how it sees.

Proof Is in the Doing

Danny Hillis, a former student of mine in AI and now a top scientist at Disney, once asked me if I had ever explained an idea to someone, only to have it misunderstood as an even better idea. Sure, I replied, it happens every time I try to explain something to Marvin Minsky (a legendary founder of the field of AI who is a professor at MIT's Media Laboratory).

Danny's point, of course, was that the inner conversation many—or all—people have when they solve problems may play the same role as a conversation with someone else. Processing thoughts expressed as word sequences must excite important thinking mechanisms buried in our language-processing hardware. Thus, the thinking lies *in* the language-processing hardware, not behind it.

I do not believe this means that understanding intelligence is a simple matter. The computations that produce intelligence are surely sophisticated and unlikely to be understood via incremental progress. My intuition is that progress based on this view will be either rapid or nonexistent; it cannot be slow. The ingredients seem to be on the table; it remains only to be seen what we can make of them. Today's reasons for optimism will either fuel a revolution or quickly prove unworthy.

Of course, critics could say they have heard all this before, citing rule-based systems or neural nets as examples of highly hyped ideas offered by overstimulated proponents of AI as the answer. Adherents of the fresh approach could respond by contrasting their views with the played-out dreams of yesterday. I prefer, however, to expend such energy on research leading to experiments and demonstrations rather than on arguments, because success has far greater force than speculation.

SCIENTIST'S NOTE:

Our Digital Future

Vinton G. Cerf

Senior Vice-President of Internet Architecture and Engineering, MCI Communications Corp.

As I look back over the last thirty to forty years of telecommunications history, I would say that the three most profound developments have been fiber optics, satellite communications, and packet switching. Fiber optics has opened up an era of digital broadband communications services, which operate at many billions of bits per second. Satellite communications have truly shrunk the globe, most vividly exemplified by CNN. And packet switching, which treats information not as continuous streams of data but as electronic postcards routed separately through the network, is about to transform television, radio, telephony, and even traditional print media. Of course, few of these profound changes could have happened without the proliferation of digital electronic devices, including computers of all kinds, which can take advantage of these communication services.

Intelligent programmable devices are increasingly finding their way into our everyday lives. My laptop is a constant companion as more and more of my daily work is done online and with the aid of software. The Internet, a child of the packet-switching revolution, expands daily and is becoming a kind of global memory where information of all kinds is accessible to and shared among the world's population. Although still in an early phase of its evolution, having been in commercial service only since 1990, the Internet seems destined to house the world's information infrastructure, facilitating research, education, electronic commerce, personal communication, government, and entertainment.

In the future it seems highly likely that bioelectronic devices will become more common, and that extraordinary prostheses such as cochlear implants and artificial retinas will emerge not only to correct deficits but perhaps even to enhance normal function. One hopes for the day when spinal cord injuries, for example, can be repaired by programmable implants that mediate nerve communication across what is today an unbridgeable gap. The time is also rapidly approaching when digital display technology will be preferred over paper for reading volumes of material. These thin devices may become part of digital glasses that project virtual images onto the retina, affording privacy, light weight, and the convenience of low-power operation.

Wearable computing devices are becoming common

in the form of two-way pagers and Internet-enabled cellular phones. Rings worn on the hand that contain the equivalent of a smart card (a small specialized computer), a Java interpreter for Web pages, and a simple input/output interface for sending and receiving information are already here. Voice recognition by computers augments keyboard interfaces. Appliances will be Internet-enabled and become part of the general computing and communications environment.

Within twenty-five years, it seems likely that we will have colonies on the moon and Mars, and in fifty years on the moons of the outer planets. These will all be bound together in an interplanetary Internet whose design is already under way. If these ideas seem far out, one can simply observe the last fifty years of advances in medicine, space, and digital technology to be persuaded that most long-range predictions, like this one, prove largely timid.

One good counterexample is language translation, which was thought to be just a matter of putting dictionaries into computers in the 1960s. The goal turns out to be much harder to reach than we thought, but even now, crude translation of World Wide Web pages from one language into another is common. Perhaps, with enough computer power and memory, real-time translation may become feasible.

As we speculate on the possibilities of our digital era, we would do well to keep in mind that the most profound value of these new technologies is the ability to communicate with friends and family, colleagues and business acquaintances more fully and reliably than ever. We are indeed becoming a networked world.

PROTECTING THE NEST

The Path of Enlightened Self-Interest: A Quarter Century of Environmental Progress

William G. Scheller

Environmental Writer

The fuel that fed the fire was unspectacular—trash, fallen tree limbs, a good lashing of heavy bunker oil. But when Ohio's Cuyahoga River went up in flames on June 22, 1969, far more was ignited than a single, badly defiled urban waterway.

The environmental movement of the past quarter century certainly cannot be traced to a single event. But the burning of the Cuyahoga was a signal, an alarm in the night. People began to wonder: if human activity can so thoroughly pollute a river as to make it catch fire, what else might we be doing to our environment?

By 1974, when *NOVA* was first broadcast, newspapers and airwaves were filled with reports, few of them heartening, from what had come to look like a battleground on which humankind struggled with its own future. Lake Erie, we learned, was in its death throes, the victim of an oxygen-depriving process called eutrophication that had been linked to the phosphates in our detergents. The National Oceanic and Atmospheric Agency reported that a 700,000-square-mile portion of the Atlantic Ocean was fouled with "significant quantities of oil, tar, and bits of plastic." Urban air quality—particularly in the Los Angeles basin, always the nation's smog bellwether—was so bad that the fledgling Environmental Protection Administration (EPA) proposed transportation controls designed to alter the American public's "long-standing and intimate relation to private automobiles."

The raw materials of pollution were likewise the raw materials of progress, and by all indications we were using them at a rate that would not only guarantee unsupportable levels of the former but also provoke a sudden, sharp end to the latter. Even before the Arab oil embargo of late 1973, President Richard Nixon was calling upon Americans to develop an energy conservation ethic. Otherwise, Nixon warned, we would soon face serious shortages.

In 1972 the Club of Rome had come to conclusions far more dire, outlining in its book *The Limits to Growth* a scenario for worldwide environmental and social calamity if the perpetual-growth economic model were to be pursued indefinitely. In 1968, Paul and Anne Ehrlich had pointed to the simple, Malthusian cause of all present and future environmental woes. The title of their book had said it all: *The Population Bomb*. From a burning Ohio river to a world that British author C.P.

This bird is one of the countless victims of the Exxon Valdez *oil spill in Alaska. NOVA probed the failures of technology in preventing, containing, and cleaning up the worst oil spill in U.S. history in "The Big Spill."*

Snow predicted would careen into famine by 1980, a clear continuum was suddenly apparent. Humanity was fouling its nest, and the worst was yet to come.

The alarms were not ignored. The first celebration of Earth Day had been held in April 1970, and that same year the EPA was given responsibility for enforcing federal environmental legislation. The phrase *federal environmental legislation* first applied to a pitifully small body of law, but by 1974 a Clean Water Act and an amended and reinforced Clean Air Act were already on the books.

Perhaps most important, people were beginning to appreciate the fact that there was an environment to protect. Today the idea seems patently obvious. But if we look back much further than twenty-five years we find that public consciousness held room for the idea of an economy, a nation-state, and a great agglomeration of sovereign individuals and their private concerns—but not for a notion of humanity's relationship to its natural surroundings that extended beyond excoriating litterbugs and heeding the anti-incendiary admonishments of Smokey the Bear. Then, almost overnight, Green political parties appeared in Europe, an agency of the U.S. government was charged with improving the quality of air and water, and consumers were willing to accept the fact that propellants in aerosol hair spray might actually be damaging Earth's ozone layer, and were eager to replace their favorite detergents with a new product called "Ecolo-G." This gratuitously named alternative to phosphate-laden products turned out to consist primarily of salt, and soon disappeared from the shelves. But the new environmental consciousness proved a good deal more authentic, and durable beyond what even the most sanguine environmentalists dared expect.

Of course, no such sea change in public awareness occurs overnight. Modern environmentalism did not spring full-blown from the burning waters of the Cuyahoga River or appear—in lieu of a view extending for more than two blocks—out of the miasma of an L.A. smog. In the United States the serious study of human interference with natural systems dates from a book entitled *Man and Nature*, published in 1864 by the Vermont congressman and diplomat George Perkins Marsh. Looking primarily at the historical record of deforestation, erosion, and soil depletion, Marsh concluded that human ingenuity had all too often been a disruptive force with regard to ecological equilibrium, and that an informed and humane application of scientific knowledge could serve a restorative purpose. Marsh was thus an early apostle of the idea of enlightened self-interest as an approach to custodianship of the natural environment. If humanity is to benefit from the world in which it finds itself, he believed, then humanity had better exercise forethought and restraint.

Marsh's near-contemporary Henry David Thoreau was the wellspring of a radically differing philosophy. Thoreau argued powerfully for considering nature not as a subject for careful management but as an entity that exists apart from human enterprise—something to be treasured for its innately un-human qualities. Marsh wanted to save the trees because we might need them later. Thoreau wanted to save the forest because we need to know it is there.

Thoreau's transcendentalist's-eye view of the natural world resonated in the wanderings and works of John Muir, who founded the Sierra Club and fought for the preservation of vast tracts of western wilderness. And Muir, in turn, exerted a significant influence on President Theodore Roosevelt and his substantial conservationist legacy. It was Roosevelt who synthesized Marsh's notions of the enlightened use of natural resources—a view shared by the great scientific forester Gifford Pinchot, appointed by T.R. to head the U.S. Forest Service—with Muir's romantic preservationism.

Much of the impetus for the nascent conservation movement came from individuals who were beginning to realize the toll that habitat destruction and unrestricted hunting were taking on wild species, particularly birds. The National Audubon Society was founded largely in response to the depredations of market hunters and a fashion industry addicted to lavishly feathered hats. And the rapid closing of the American frontier led to the first calls for careful conservation of Western land resources that had long been treated as inexhaustible. John Wesley Powell, the Colorado River explorer and founding member of the National Geographic Society, brought to his job as first director of the U.S. Geological Survey a commitment to enlightened land management. The federal government designated the first wilderness areas on public lands in the early 1920s.

The Thoreauvian idea of preserving wild lands for their own sake picked up momentum in the writings of Robert Marshall, forester and cofounder of the Wilderness Society, who warned in 1930 of "the tyrannical ambition of civilization to conquer every niche on the whole earth." Aldo Leopold, another Wilderness Society founder, wrote in *A Sand County Almanac* of a "land ethic" characterized by his belief that "a thing is right when it tends to preserve the integrity, stability, and beauty of the biotic community." Leopold's insistence on the interconnectedness of natural systems was an important step to a more all-encompassing environmental ethic.

That transformation, from a conservationism identified largely with intellectual elites to widespread environmental awareness, took a giant stride with the publication in 1962 of Rachel Carson's *Silent Spring*. It was Carson who first popularized the idea that what we were putting into the environment was as grave a concern as what we were taking out. Carson's "silent spring" was, of course, a not-too-distant future without birdsong. The threat to avian species was the wholesale contamination of the environment with chemical pesticides, especially DDT, and her implicit warning was that humans were ultimately as susceptible as animals to the health hazards of contaminants we introduced into air, water, and soil—ironically, in an attempt to improve our environmental circumstances by destroying insect pests.

The notion that the cure might be worse than the disease proved difficult to swallow. DDT was outlawed only over considerable objection in the United States and remains an insecticide of choice throughout much of the world. Even where it has been banned, we have learned that the harm it does—like the damage caused by some other pesticides and herbicides—long outlasts it.

It proved easier to rally the public against air and water pollution than against pesticides, but rivers, lakes, ocean, and atmosphere were subject to a grim cost accounting. Soiled skies and waters had traditionally been taken as a sign of progress, a token of full employment, with cleanup put off until the cost of a badly begrimed public image appeared to exceed the benefits of the status quo. Pittsburgh, long derided for having to turn on its streetlights at noon to dispel its smokestack gloom, had commenced a multibillion-dollar cleanup in the 1940s. But as late as 1973 car manufacturers warned that the emission-control devices needed to meet the EPA's three-year air-pollution-abatement target would add $275 to the price of a car. Later that same year General Motors announced the first successful catalytic converter, a device that, along with other advances such as electronic ignition, did indeed send car prices higher.

Clearly, however, modest increases have not stopped consumers from buying cars.

Prosperity and Protest

A DDT ban here, a catalytic converter there . . . it still doesn't add up to a world-changing shift in environmental consciousness. What was it, finally, that made the past twenty-five years such a watershed in the history of ecological awareness?

In the United States the environmental movement was the unintended beneficiary of the antiestablishment spirit nourished by years of protest against the Vietnam War. Early opposition to the war was purely political, but the culture of protest soon assumed a holistic perspective. The belligerence of the military-industrial complex against a human foe came to be identified with belligerence against nature, against a natural harmony inimical to material aggrandizement.

There were radical and liberal strains within the nascent environmental movement. At one end of the spectrum were the course correctors, still looking to fix what was wrong within the existing economic and political system. At the other stood those who argued that business as usual, even if pursued with the best of intentions, would keep humanity at odds with nature. In simplest terms, the continuum ran from catalytic converters to better public transportation to visions of neo-medieval villages built around windmills and bicycle paths. But for the first time, nearly everyone seemed to have some perspective on environmental betterment, and all the ideas had something to do with abandoning blind trust in the institutions that made things and ran them.

Another powerful impetus toward heightened environmental consciousness had to do with a long period of prosperity for First World nations. It is difficult to quarrel with belching smokestacks, or with chemical effluents in waterways, during or immediately following an economic depression. A healthy environment is hardly a luxury—but that doesn't keep it from looking like one when times are bad. The United States was in an expansive mood a quarter century ago, and environmental quality was something the middle class felt it was time to shop for and spend money on.

That expansive mood, needless to say, quickly turned sour when Arab nations imposed an oil embargo in the aftermath of the Arab-Israeli Six-Day War, and the words "energy crisis" were suddenly on everyone's lips. Energy shortages led not only to gasoline lines but to a sharp round of "stagflation," the newly minted term for a paradoxical mix of inflation and recession. The movement toward improved environmental quality and energy conservation, though, did not evaporate with the good times. In large part because environmental depredation was now linked in the public mind with an out-of-control energy budget, enthusiasm for conservation and cleanup prevailed. That enthusiasm would be threatened in the far more politically conservative environment of the 1980s, when traditional business values were ascendant, but by then many of the more progressive environmental initiatives had gone mainstream. Ronald Reagan's Morning in America might mean trimmed EPA budgets and congressional battles over adding to the endangered-species list, but no one seriously advocated removing scrubbers from smokestacks or knocking down sewage-treatment plants.

Finally, there was a symbol—and who knows what a symbol may count for?—of Earth in all its beauty and fragility. Ever since *Apollo 8* trained its cameras on the planet, we have been able as never before to look at the place where we live and realize that it is all we have and all we are going to get for any foreseeable future. The photograph was taken from too great a distance to tell if any rivers were burning, or how many lakes had been scoured clean of life by acid rain, or where the toxic waste dumps were hidden. But it looked as if any of those things would have been a terrible shame.

The Good News

Having regarded that blue ball in space—having absorbed the lessons of the past, recoiled at the depredations of the present, and filtered it all through a new perspective that demanded we Do Something—just what was it that we did (and didn't) do?

The broad list of concerns first seriously discussed during the past couple of decades includes sewage treatment, overburdened landfills, recycling of metals, glass, and paper, development of alternative energy sources, preserving wetlands, pollution of groundwater supplies, biodiversity, the plight of tropical rainforests, acid rain, deterioration of Earth's ozone layer, and global warming. None of these problems have lent themselves to easy solutions; a number will be around for decades to come. But admitting that they exist, in many cases, has been as hard as settling down to solve them.

As for results, the Cuyahoga River, at least, is no longer in danger of catching fire. During the decade following the 1969 blaze, oil spills on the river fell by more than 90 percent, Cleveland's steel mills voluntarily stopped using the Cuyahoga as a drain for cyanide-contaminated water, and chemical dumping dropped substantially. By 1977 ducks lived on the river. These were no isolated accomplishments; they have been repeated throughout the United States and the developed world. Public and private U.S. expenditures on pollution abatement and control, in constant 1987 dollars, rose from less than $50 billion in 1973 to more than $91 billion in 1993, the last year for which figures have been published. Quantities of contaminants such as phosphorus, lead, and cadmium in American rivers and streams fell precipi-

Christopher McLeod/Earth Image Films

Near Leadville, Colorado, abandoned mining wastes have contaminated water with heavy metals. In "Poison in the Rockies," NOVA reported on water pollution in the West.

tously over the same period. In 1973 more than 11,000 oil-polluting incidents involving some 15 million gallons were reported in and around U.S. waters. In 1993 roughly 9,700 incidents accounted for only 1.5 million gallons.

Our air is undeniably cleaner. Los Angeles failed to meet federal air-quality standards on 226 days in 1986; by 1994 the figure was down to 136. Nationally, emissions of particulate matter, sulfur dioxide, volatile organic compounds, carbon monoxide, and especially lead—now banned from motor fuels—have all fallen, even though the number of vehicle-miles traveled per year has doubled, and the number of passenger cars has grown from 80 to 120 million. The key factors have been automobile gas mileage, which improved 62 percent between 1970 and 1994, and the universal application of the catalytic converter.

Recycling has proven another promising trend. In 1970 only 8.6 million tons of recyclable materials were recovered in the United States; by 1994 that figure had climbed to 52 million tons. Recycling and other environmentally related jobs now employ some 1.3 million Americans, roughly three times the number engaged in those fields twenty years ago.

The past two decades have witnessed notable successes in expanding the range and numbers of high-profile endangered species. Wolves roam again in Yellowstone, many marine mammals are protected, the trade in elephant ivory has been heavily suppressed, and, with DDT gone, peregrine falcons are making a comeback. Americans have yet to face a silent spring.

In 1972, the last houses were built on New York State's Love Canal site; six years later, the federal government began buying out Love Canal residents, and President Carter declared a federal emergency at the former chemical waste dump—the first ever associated with a human-made environmental disaster. In 1980 Congress authorized a Superfund for cleaning up hazardous waste sites, often characterized by the presence of heavy metals, dioxins (a cancer-linked byproduct of industrial processes such as paper bleaching), and highly toxic polychlorinated biphenyls (PCBs). By 1994 the Superfund list included twelve hundred sites, although progress on cleanup has been slow and thousands of other sites remain a legacy of a less-enlightened era. Still, much more stringent codes now apply to discarding the approximately 60 million tons (1994 figures) of U.S. waste classified as hazardous each year. Similar strictures are enforced throughout most of the developed world.

In 1973, who knew or cared how much energy was consumed each year by a refrigerator or water heater? Today appliances wear energy-efficiency labels as visible as their price tags, and all of them use power far more sparingly than the machines sold twenty-five years ago. Insulation is as unquestioned a building component as indoor plumbing; many homeowners in areas of climactic extremes spout R-values in casual conversation.

Among the greatest accomplishments of this quarter century are the things that *didn't* happen. Despite severe localized famines often attributable to sub-Saharan desertification, regional crop failures, and breakdowns in food distribution caused by war or civil unrest, C.P. Snow's 1968 prediction that the world would soon be awash in a "sea of hunger" has not come to fruition. The Green Revolution that improved crop strains and farming techniques has yielded astounding results. India, which produced twelve million metric tons of wheat on fourteen million hectares in 1964, harvested fifty-seven million tons on twenty-four million hectares in 1993. Indian grain reserves that year totaled twenty-five million metric tons—while a century before, with only a third of its present-day population, the subcontinent experienced a famine that took ten million lives.

What's more, we haven't run out of oil or vital minerals. In fact, in 1990 the cornucopian (opposite of Malthusian) economist Julian Simon won a bet he had made a decade earlier with Paul Ehrlich regarding changes in the prices of copper, chrome, nickel, tin, and tungsten. Ehrlich had predicted that scarcity would drive prices for all five through the ceiling—but in fact they all fell, thanks to more efficient use, recycling, and the development of substitutes. Simon cited his victory as proof of the triumph of human ingenuity, while Ehrlich conceded only that his calculations of the Malthusian timeline were off.

Deep Ecology versus Human Ingenuity

No one without a highly suspect agenda would argue that all the environmental news of the twentieth century's closing quarter has been good. In terms of global warming, perhaps our era's most important ecological issue, the U.S. Congress has yet to ratify the most recent treaty, drawn up at an 1997 international conference in Kyoto, governing greenhouse gas emissions.

The question of whether carbon dioxide from fossil-fuel combustion has trapped solar heat near the Earth's surface has been taken seriously since the late 1970s, when atmospheric scientists first raised the idea of a relationship between CO_2 levels and rising air temperatures (and sea levels). The fact that the 1980s were the warmest decade since record keeping began added urgency to the debate and prompted European nations and Japan to set goals for stabilizing and eventually reducing carbon dioxide emissions. Arguing for a more cautious program are the U.S., where industrialists voice concern about the economic impact of cutting emissions from power plants and other processes (remember car makers' warnings about the costs of pollution-control devices?), and developing nations, who fear their economies may be hobbled by nations that have already contributed more than their share to global warming. A notable exception are low-lying island and seacoast nations, which have the most to lose from rising seas.

Depletion of the Earth's protective ozone layer remains significant, despite developed nations' ban on chlorofluorocarbons (CFCs) and a three-quarter reduction in U.S. output over the past ten years. The ozone shield, the component of the stratosphere that protects the Earth's surface from harmful ultraviolet rays, has seen a 2 to 3 percent decline worldwide; in a seasonal hole above the Antarctic, as much as half of the normal portion of ozone disappears. Solar ultraviolet radiation is associated not only with skin cancer in humans but with negative effects on the plankton that feed the bottom of the ocean food chain. Ozone loss has been implicated in the alarming rise in deformities among frogs, which may have reached 45 percent in parts of the United States.

Emissions of nitrogen oxide, a byproduct of fossil-fuel combustion that irritates the human respiratory tract and is toxic to many plant species, are stuck at early 1970s levels. Although catalytic converters have cut the amount of nitrogen oxide that vehicles contribute to ground-level ozone, a major component of smog, stationary sources such as power plants have yet to be effectively controlled.

Deforestation, particularly in the Amazon basin, persists despite worldwide awareness of its impact on weather patterns and the extinction of perhaps millions of species, many still unidentified and none as glamorous as the bald eagle, wolf, tiger, or African elephant. The seas are still recovering from years of dumping of sewage, ships' waste, and dredged material that frequently contains toxic heavy metals; nations not party to the London Dumping Convention of 1975 and its subsequent protocols continue to foul the waters. Much of the world's saltwater resource is also severely overfished. Catches of traditionally plentiful species such as Atlantic cod, Pacific and Atlantic salmon, halibut, and pollock have all declined precipitously. Development worldwide continues to fill wetlands, eliminating not only habitat for birds and animals but a cornerstone of the marine food chain and a natural system of flood control.

The list of unfinished business goes on. Because oil has stayed cheap, research into nonpolluting energy sources heralded in the 1970s—solar, wind, geothermal, and other technologies—has lagged. Uncontrolled urban sprawl continues to devour land, particularly in the developing world. And in formerly communist nations, massive cleanup efforts are required to reverse the damage done by state-managed economies whose disregard for environmental quality and human health rivaled that of laissez-faire capitalism at its worst.

And what about population, the biggest bugbear of all? For as long as there has been an environmental movement, the geometric expansion of world population has been perceived as a driving force behind the world's twin dilemmas of pollution and resource depletion. The argument the Ehrlichs made so cogently in *The Population Bomb* was a difficult one to refute, given the stark data on which the book rested. Between 1750 and 1950 world population grew from 1 billion to 2.5 billion. From 1950 to the present the total climbed to nearly 6 billion. The conclusion was inescapable: with population doubling time shrinking at a harrowing rate, and the gentle upward arc of fecundity turning into a vertical line, humanity appeared to be heading for a population crash mercilessly played out through famine, disease, and war—it was anyone's bet which apocalyptic horseman would be the first to arrive.

But without debating whether significant portions of the world are severely overpopulated—the sprawling cities of the developing world come first to mind—we can now look at newer projections and conclude that the population explosion will quite likely resemble the spike predicted by some futurists in the 1960s. The reason for such optimism is that birthrates are declining in nearly every part of the world. In the early 1950s the global fertility rate was 5 births per woman—contrasted with the replacement (or zero population growth) rate of 2.1. By 1997 the rate was 2.8 and still falling. In the developing world the rate has dropped from 6 to 3 births per woman since 1970; in Bangladesh, for example, it's fallen from 6.2 to 3.4 during just the past ten years. (The U.S. rate has averaged 1.9 since 1972.)

None of this means that the world's population is no longer growing. It is, simply because formerly high birthrates have supplied large numbers of women who are now of childbearing age. What it does mean is that barring a drastic turnaround in fertility rates, population is not going to increase exponentially to the point of catastrophe. United Nations statisticians, in fact, are revising downward their estimates of global population fifty years from now. Instead of reaching 10 billion or more by 2050, the total may well top out at 8.5 billion and then start to decline.

These are encouraging figures, although they are hardly cause for complacency. For one thing, we must still consider the half-century of growth ahead; in China alone, recent projections call for a population increase of 400 million—from 1.2 to 1.6 billion—before the numbers begin to decline around 2050. The most daunting set of problems, though, has to do not with raw population growth but with economic growth and improving the standard of living in developing nations.

The situation can be simply defined: the typical American family has some forty times the environmental impact of an Indian family, and one hundred times the impact of a family living in Kenya. The ecological dangers inherent in raising the living standards of less-developed countries to those of the U.S., Europe, or Japan are obvious, yet no one with an ounce of compassion would suggest subverting development in areas where suffering, not mere inconvenience, is the daily lot of millions.

The conundrum we face is that increased prosperity drives down birth rates but also drives up levels of pollution and resource extraction. The challenge is for developed nations to transfer to their poorer counterparts the technological instruments that will enable them to bypass the messier stages of the Industrial Revolution—with all of their attendant waste and pollution—on the path to improved living standards.

Energy development is the first priority. Aside from

the question of whether there are enough fossil fuels, discovered or undiscovered, to power billions of additional First World lifestyles, the extraction and consumption of these resources would exert an apocalyptic toll on the environment. The alternative to burning fossil fuels is to promote renewable, sustainable energy sources, perhaps chief among them solar generation of electricity. Photovoltaic technology, after all, is splendidly suited to many of the resource-poor, sunlight-rich places that most need electric power for economic development.

Another imperative is to help implement land-use and transportation schemes designed, to a far greater extent than our own, around public conveyance rather than individual automobiles. Lessons we have already learned, if not fully implemented, about recycling, substituting renewable for nonrenewable resources, and reducing waste streams must also be part of the technology-transfer picture.

All these prescriptions for the have-not nations prompt the question of what people ought to be doing on the more prosperous side of the fence. Depending on which of our environmentalist traditions we listen to, the answer is either a more vigorous pursuit of the policies that have led to limited but real success since the early 1970s or a radical revision in our thinking about what it means to live on Earth. The latter is the position of proponents of what has been called "deep ecology," a worldview that holds the planet to be a self-regulating mechanism in which technological humankind is an invariably destructive intruder.

But deep ecology faces two insurmountable problems. First, any prescriptive philosophy based primarily on anti-materialism will invariably run afoul of the vast majority of human beings, who cannot and will not renounce the physical, acquisitive side of their natures. Like the Jansenist theologians of the seventeenth century, deep ecologists—heirs, in part, to the tradition of Thoreau—believe that humanity is inherently corrupt. Because this notion will never attract many adherents, it is doomed from the start. So are we all, if the deep ecologists are correct.

The second problem with the radical viewpoint is that it sidesteps the basic fact that humanity is a part of nature. So is human ingenuity, which leads us to hope, and to act on the idea that whatever messes we can get ourselves into we can get ourselves out of. A mystical, essentially irrational view of nature has its purposes; after all, the romantic instinct leads us to preserve millions of acres of wilderness most of us will never see. But concern for people's material well-being is what must ultimately drive any worldwide effort to preserve an environment in which life is worth living. This is the path of enlightened self-interest, the only one we can realistically follow. The accomplishments of the past twenty-five years, however halting and incomplete, are proof of our ability to eventually realize what's good for us. So is the braking mechanism that has begun to work on world population.

That doesn't mean that we do not have some serious motivational problems to overcome. Another human attribute, along with inventiveness, is a fondness for the path of least resistance: we aren't given to taking bitter medicine, except in times of crisis. This problem is compounded by the fact that recent history suggests that any foreseeable environmental progress must occur within the context of market economics, a system—however admirably flexible—that favors short-term gains. We face the challenge of making markets respond to long-term problems, and making them give us—accompanied by the requisite profit—the clean energy technologies, sustainable agriculture, open land, biological diversity, and livable cities that we want and need.

The idea of a steady-state economy has long been anathema to traditional economists, who see constant growth as intrinsic to healthy markets. But if population projections for the next century are correct, we will sooner or later have to come to grips with a consumer base that is stabilizing and in many places actually shrinking. A steady-state economy might thus be mandated by sheer numbers, but it doesn't have to be stagnant. It might well be an economy that grows by plowing back over its old fields, doing things right that were formerly done merely in haste in an attempt to chase ever-expanding markets.

Look at the growth of employment in the environmental sector. At every turn in our recent history the very steps initially lamented as job killers and economy busters have resulted in new opportunities for employment and profit. There are millions of solar photovoltaic roofs to be constructed, thousands of miles of railroads to be rebuilt, hundreds of thousands of tons of waste to be recycled. There is plenty of work to be done creating new materials and industrial processes that will supplant the polluting, resource-extracting technologies of the industrial age.

There is far more opportunity to be seized in cleaning our rivers, we are beginning to learn, than in setting them on fire.

SCIENTIST'S NOTE:

Forecasting Tornadoes

Howard B. Bluestein

Professor of Meteorology, University of Oklahoma

It has been said that tornadoes are the last frontier of meteorology. Magnificent to watch yet lasting briefly and affecting only a small area, they resist observation, let alone complete understanding.

In 1973 storm chasers in Oklahoma got lucky: photographs and 16mm movie footage taken on the fly documented the complete life cycle of a powerful tornado, while a Doppler radar 50 kilometers away probed its parent "supercell" storm. The radar showed that a narrow column of rapidly rotating air at high altitude preceded the appearance of the tornado on the ground by twenty to thirty minutes.

Although our understanding of tornadoes has improved in the ensuing twenty-five years, we still do not know exactly why they form, nor can we distinguish supercells that spawn tornadoes from those that do not. The main obstacle remains the difficulty of ensnaring tornadoes in our observational net and making

NOVA showed the electrifying spectacle of scientists provoking bolts from above to learn more about nature's fleeting rivers of electricity in "Lightning!"

detailed measurements amid such violent and capricious events.

In the early 1980s meteorologists attempted to make such measurements by placing a device named TOTO—the Totable Tornado Observatory—in the expected path of oncoming tornadoes. But because the storms dissipate or change direction so quickly, the device largely failed to capture useful information. When computer simulations around the same time suggested that buoyancy and wind shear determine whether a storm can generate a rotating column of air, we began to release instrumented

AP/Wide World Photos

A six-story building that collapsed in the aftermath of the earthquake that rocked Kobe, Japan, on January 17, 1995. A year earlier, a quake of similar magnitude struck the Los Angeles area. NOVA looked at what scientists learned from the twin calamities in "The Day the Earth Shook."

balloons high in the atmosphere to record the thermodynamic environment in which storms grow.

Not till the late 1980s did portable (although still heavy) Doppler radar appear, enabling scientists to set up quickly just a few miles from a sighted tornado. By the summer of 1991 we had verified wind speeds in excess of 261 miles per hour. Although our relative lack of mobility still limited our ability to measure a tornado's entire life span on the ground, airborne Doppler radar mapped out the wind field in supercell storms that same year. By 1995 a number of tornadoes had formed under its watchful eye.

Analysis of this information reveals that tornadoes form on even shorter time and space scales than we can yet resolve. To overcome this, meteorologists are attempting to use two high-power truck-mounted Doppler radars to map parent vortices as well as larger tornadoes themselves. Meanwhile my group is experimenting with even higher-frequency, higher-resolution yet smaller truck-mounted radar to map the wind field inside tornadoes.

In the future instrumented pilotless aircraft, rocket probes, and Doppler laser radars, which can sense motion in air without even cloud droplets or precipitation and might be mounted on helicopters or other airborne platforms, could move along with the tornado. From these observations and detailed computerized simulations, we should attain a good understanding over the next twenty-five years of how and why tornadoes form in supercells. Our ability to predict them, however, may still be limited by the difficulty of measuring and predicting the state of the atmosphere at every point at every instant. Sensitivity to initial conditions may seriously constrain the time over which we can make pinpoint tornado forecasts.

SCIENTIST'S NOTE:

The Mystery of Snow

Othmar Buser

Swiss Federal Institute for Snow and Avalanche Research

Avalanches—so graceful and lithe, so full of wonders that one might forget that their destructive force can be fatal—are obviously made of snow, but why is there snow in the first place?

Water vapor transforms into snow crystals of a myriad different patterns, all hexagonal. Why? What for? What a waste! Wouldn't one form do? We may know *how* crystals form given atmospheric conditions, but *why* they are as they are we scientists do not know. (The priests may know, or the philosophers.) Nevertheless it is wonderful to see and enjoy such a splendor—to see the connections that are visible but also to know that the essential is invisible to the eye.

When I entered the Swiss Federal Institute for Snow and Avalanche Research more than thirty years ago, I decided to do serious work. However, I already knew that "the essential is invisible to the eyes. One does not see well but with the heart" (see Antoine de Saint Exupéry, *Le Petit Prince*). I was happy to work with snow because it is inorganic, behaving according to basic physical laws, yet it is also a link to organic life: melting snow becomes water, and without water, there is no life.

The most impressive story I know that reveals the invisible side of snow is the German fairy tale "Frau Holle." In this tale a diligent, open-minded maid was walking through the spring, summer, and autumn seasons and arrived at last at a house where an old woman with frightful teeth looked out the window. The girl was terrified at the sight of the woman, but when she offered her house for the winter, the girl accepted. She could stay for free, except that she would have to toss the feather quilts such that the down came out. Then, the old lady explained, snow would fall on the earth. From this fairy tale German children know *why* there is snow, and they enjoy it.

For me there is still another aspect to the story: the snow covering the soil and protecting it from excessive frost also forms avalanches—the frightful teeth of the old woman. For people living in areas with deep snow it was more than the inorganic porous medium it has become for us. Snow was alive, sort of a wild animal. When you disturbed it, it woke up. An avalanche was a beast, but they could live with it.

Near where I live is a hamlet in an avalanche runout zone. Settlers who erected their dwellings there centuries ago were aware of the threat. To protect their homes, they decided to build a church some distance up the slope. God would not be so stupid as to destroy His own house, they argued, but he could hardly mind if they helped Him. So they added an avalanche wedge in back of the church. (Such a wedge, made of stone, concrete or earth, points up toward the avalanche, splitting a moving snow stream so that it flows past the building.) It served the purpose. This attitude is characteristic of how our ancestors dealt with natural hazard. Help yourself and God will help you.

Avalanches more and more lose their enigma because of nice and much-appreciated work—serious efforts by even more serious people (see *Le Petit Prince*, part I, concerning serious people). We think we have tamed the beast, if not exterminated it. But trying to see with my heart I think we may sometimes forget the essential.

LEAVING HOME

The Incredible Trip:
A Quarter Century of Space Science

Richard Maurer

Science Writer

In 1974, space exploration was in a funk. The Apollo program had wrapped up two years earlier after putting a total of twelve astronauts on the Moon. The three-person *Skylab* space station was hosting its third and final crew. The Soviets were forging ahead with their own space station, called *Salyut*, but it paled in size and sophistication to *Skylab*, which was then the largest object ever launched into Earth orbit (and within a few years would be the largest object ever to fall from Earth orbit). Robotic missions, those poor cousins of the human spaceflight program, had sent back a tantalizing stream of data from Mercury, Venus, Mars, and even Jupiter, and more missions were in the works, but the U.S. and the Soviet Union had experienced high failure rates on such flights and everyone was wary of expecting too much.

Most significant of all, the space race was over. The starting gun had been *Sputnik 1* in 1957, when the Soviets surprised the world by launching the first artificial satellite. They followed with a string of audacious missions that left America in the dust as its own space efforts mostly fizzled or exploded on the launch pad. In 1959 the Soviets were the first to hit the Moon when they crash-landed a probe there. In 1961 a U.S. experiment to send a human on a brief suborbital ride was upstaged by Yuri Gagarin's spectacular flight orbiting the globe. Shortly thereafter, President John F. Kennedy announced America's own audacious plan: to put a man on the Moon by decade's end. No one doubted that the Soviets would try to get there first. They did try and were the first to reach a number of milestones, such as launching a multi-person crew, docking ships in space, and landing a robotic craft on the Moon. But they faded in the final stretch as America swept the board with its mammoth Apollo effort. Preempted in the feat to end all feats, the Soviets decided not to attempt their own piloted lunar adventure and took to sending automatic probes to poke around the lunar surface instead—a pale substitute, it seemed, for having real people on the spot.

Yes, the space race was a tough act to follow. But in retrospect the push to explore wasn't really slowing down, it was just switching contestants. People in space suits were about to hand the baton to a new breed of instruments—machines that are still leading the way today.

The launch of *Sputnik* created a rush to answer long-standing questions about the space frontier: What was the nature of the space medium? What did the far side of the Moon look like? Did Venus have an Earth-like at-

NASA

Astronaut Charles "Pete" Conrad holds a sample of lunar soil at the Apollo 12 *landing site on the moon in 1969. In "Twenty-Five Years in Space,"* NOVA *reviewed the impact of the space age on American culture.*

mosphere? What exactly *were* the "canals" on Mars reported by some astronomers? And above all, could humans survive in space?

The public relations aspect of the space race put a premium on flights with people, first to show that they could be done and then to explore just *what* could be done, a quest that culminated with the Moon flights. But by 1974 the simple questions had been answered and more challenging mysteries beckoned: Was there life on Mars, or was that prospect as illusory as the canals had proved to be? What processes were at work on Venus, where spacecraft had recorded temperatures hotter than a self-cleaning oven? What did Jupiter's mysterious moons look like? What about Saturn's rings? Answering such questions was an attainable objective for robots, but where did that leave humans? There had never been a practical reason to send people into space, and there was even less reason now. For astronauts, the only place to go after the Moon was Mars, hundreds of times more distant and prohibitively expensive to reach. The U.S. space agency, NASA, therefore decided to refocus its piloted efforts on developing a reusable rocket—a space shuttle—designed to provide inexpensive access to low-Earth orbit, the shallows of the space frontier, which would serve as a staging area for future human endeavors such as space stations and, in time, Mars missions. For their part, the Soviets chose to concentrate on their existing space station, *Salyut*,[1] along with their tried-and-true expendable rockets. Meanwhile the job of traveling millions and even billions of miles to actually explore space would proceed with robots.

The Inner Planets

Unfortunately, things did not bode well for the robots. In 1974 four Soviet probes—two orbiters and two landers—arrived at Mars. All but one orbiter failed. The same year a test of the new rocket that would launch the U.S. Viking missions to Mars in 1975 also failed. Long

NASA

In 1976, the NASA Viking lander sent back this picture of the Martian landscape—the culmination of three centuries of speculation, investigation, and discovery about the Red Planet.

reconciled to such setbacks, reporters credited the disasters to a Great Galactic Ghoul, a fiend that thwarted earthly efforts to explore deep space. The Ghoul seemed to reserve special protection for the Martian surface, since four out of four attempts to land—all by the Soviets—had met with disaster.

Viking would try to break the curse. Conceived as an all-out assault on another world in the no-expense-spared style of Apollo, Viking nevertheless cost only 4 percent as much as the piloted Moon landing program—a disparity that underscored the greater efficiency of robotic exploration. Using the most powerful rockets then available (which worked this time), Vikings *1* and *2* set forth without mishap in late summer of 1975. Each vehicle was composed of a sophisticated orbiter and an even more sophisticated lander. The two flotillas were an insurance policy against failures of the sort that had plagued two out of six previous U.S. Mars missions (none were landing attempts). Barring Ghouls or other obstacles, *Viking 1* would snap the first picture of the surface of the mysterious red planet—a historic moment programmed to take place on the U.S. bicentennial: July 4, 1976.

As it happened, the extraterrestrial celebration had to be delayed, but therein lies a tale about the new adaptability of robots teamed with their human controllers on Earth. *Viking 1* arrived in Martian orbit on June 19, 1976, ahead of *Viking 2* by more than a month. Controllers had already selected a safe, flat landing site after consulting photographs from the *Mariner 9* mission, which had orbited Mars in 1971–72. However, scientists were shocked when they received Viking's preliminary orbital reconnaissance of the site, which showed steep cliffs pockmarked everywhere with small craters—details that had eluded *Mariner 9*'s cameras. With the orbiter and lander still attached, *Viking 1* went into explorer mode as controllers reprogrammed the mission to search out a safer destination. Many days later a site was found, and the lander descended to a perfect touchdown on another noteworthy anniversary: July 20, 1976—seven years to the day after humans had first walked on the Moon.

Viking's headline experiment was a miniature laboratory designed to test for signs of life. Had such evidence turned up, it would have been big news indeed. Instead, the tests were ambiguous at best, and one of the mission's highlights turned out to be something far more subtle: photo after photo of arid but Earth-like terrain. Unlike the images made on the lunar surface, which had a stark, alien quality owing to the utter lack of an atmosphere, the Mars photos revealed a place one could step into... imaginatively speaking. It was easy to picture oneself on the red planet, which resembled parts of Utah. This was less true of the thousands of images made from Martian orbit, which showed an uninviting world of ancient crater fields, desiccated river beds, vast rift valleys, and titanic stone-dead volcanoes, but enough about Mars was reminiscent of Earth to make it conceivable that life once flourished there. Like photos in a vacation brochure, *Viking*'s images revealed a place one could dream about.

Viking 2 was likewise successful and together the two missions compiled a body of data that have allowed scientists to continue exploring Mars right here on Earth. Measurements by the landers showed that the Martian atmosphere has a distinctive mix of gases unique among any known planets. When researchers discovered the same mix in a dozen or so meteorites found in different places on Earth, they knew they had in hand actual pieces of Mars that somehow had been blasted off the planet and made their way here.[2] Thus one of the long-term objectives of planetary science—obtaining samples of other worlds—was met more easily and cheaply than anticipated. Some meteorites are now known to be from the Moon, and countless others are certainly pieces of various asteroids. (Scientists nevertheless plan to send spacecraft to retrieve direct samples of Mars, asteroids, comets, and other bodies to obtain well-documented, uncontaminated specimens.)

Viking provided such a glut of Mars data that it put flights to the red planet out of business for a time. Another expedition was not mounted until 1988, when the Soviets launched a pair of probes to explore the Martian moon Phobos. Unfortunately, the Ghoul got them. In 1993 the U.S. returned to Mars with an orbiter designed to map the planet's geology. The Ghoul struck again. Not until 1997 did a successful probe revisit the red planet. This mission, called *Mars Pathfinder*, provides a good illustration of how much robotic technology has improved in the interim. If Viking cost only 4 percent of the Apollo budget, then *Mars Pathfinder* is a bargain indeed, for it cost a mere 9 percent of Viking (in inflation-adjusted dollars)—meaning that hundreds of such missions can be mounted for the price of Apollo. Even considering that *Pathfinder* is a single probe compared with Viking's four, it is still about a third as expensive per spacecraft as Viking.

Why so? Although rocket technology has changed little in the two decades since Viking, robots have gained ground by leaps and bounds. They can do the same tasks—and more—with less weight and fewer dollars. Savings come not just from increasingly capable computers but also from innovative engineering. Instead of relying on bulky retro-rockets, for example, *Pathfinder* took a tip from Detroit and deployed airbags an instant before impact. After the lander took a few harmless bounces across the red planet, the bags deflated and *Pathfinder* went to work. Savings also stem from using fewer people to staff the mission and above all from careful management to keep costs down, in line with achieving NASA's goal of sending smaller, cheaper, and more frequent flights to the planets.

Pathfinder's miniature rover explored a debris field from an ancient flood that once scoured Mars with impressive force. There is no possibility of liquid water on the surface of Mars today because of low atmospheric pressure, so evidence that great floods once inundated the planet raises several questions: Where did the water come from? Where did it go? Did life evolve while it was around—and if so, is it still there? No one yet knows the answers.

Arriving at Mars a few months after *Pathfinder* was another NASA spacecraft: *Mars Global Surveyor*, an orbiter on a high-resolution mapping mission. And in the years ahead a couple of probes will appear at Mars every other year when planetary positions are most favorable.

Mounted by the U.S., Russia, Japan, and the European Space Agency (ESA),[3] these missions are using Mars as a test case to explore a world that is both like and unlike Earth, shedding light on why Earth is the way it is, how life got a foothold here, and whether it has also become established elsewhere.

Venus shines a different light on our earthly situation. The evening star actually comes closer to Earth than Mars and is nearly the same size as Earth (Mars is only half the diameter of Earth). The first probe ever to photograph the surface of another planet landed on Venus in 1975, took a quick picture, and then expired in the 900-degree heat. The mission, *Venera 9*, was one of a long line of Soviet flights to Venus, of which about half have been successful. Venus is perpetually sheathed in thick carbon dioxide clouds, trapping solar radiation and producing a greenhouse effect run amok. *Venera 9* looked out on this hellish scene and recorded an eroded field of volcanic rocks from the foot of the spacecraft all the way to the hazy horizon. An equally bleak landscape greeted *Venera 10*, which set down twelve hundred miles away a few days later. Since then four more Veneras have landed and photographed a world of unrelieved desolation. (Mars is no less barren, but somehow its sharp shadows and gentle dunes make it seem more agreeable.)

Because clouds continually hide the surface of Venus, the only way to get a global view is to map the planet with radar. Four Venus orbiters have undertaken such missions, the most recent being *Magellan*, a U.S. probe that operated from 1990-94 and produced a detailed map of virtually the entire planet at a resolution of 240 feet. Given the weird topography that emerged—which appears to have been carved by molten lava to the same extent that Earth has been shaped by running water—one would expect to hear that scientists believe Venus to be the most alien place yet visited by spacecraft. In fact, they point out that Venus bears a striking resemblance to Earth. Both are geologically active planets with young surfaces and atmospheres that are continually recycling various gases. It could be that Venus is what Earth will become a billion years from now, when the sun brightens, the oceans evaporate, and our climate heads into a runaway greenhouse state. Billions of years in the past, when the sun was cooler, Venus was perhaps like Earth is today, with oceans and possibly even life. Proof of such speculation must await future missions—although Venus has yet to become as intense an object of public interest and hence funding as Mars.

The Outer Planets

If the past quarter century of space exploration has seen anything as momentous as the earthly exploits of Lewis and Clark, it is the missions of Voyagers *1* and *2*. These Viking-era spacecraft left Earth in 1977 with Jupiter as their first stop. Jupiter had already been probed in the early 1970s by Pioneers *10* and *11*, rudimentary machines that were designed to prove the technology and techniques of deep space travel. At that time no one knew if the asteroid belt between Mars and Jupiter was passable or if delicate spacecraft electronics could withstand Jupiter's intense magnetic field. Both perils proved survivable, and the Pioneers returned intriguing data about radiation, charged particles, and other aspects of the space medium around the solar system's largest planet and its many moons.

When the Voyagers reached Jupiter in March and July of 1979, they sent back astonishing views of the planet's many-hued, stormy atmosphere. But their greatest achievement was revealing the bizarre world of the moons. Nothing like them had ever been seen. Io is convulsed by constant volcanic eruptions. Europa has a maze of strange lineaments resembling cracked ice floes and may harbor an ocean of liquid water beneath its frozen crust. (If true, this would make it the only solar system body besides Earth known to have water in a liquid state, raising the possibility that life may exist there in some form.) Ganymede has a corrugated appearance that may be due to deformation of its icy crust. Callisto is the most heavily cratered solar system body yet observed—quite a distinction considering the pummeled faces of the Moon and Mercury. The Voyagers also discovered three new satellites (bringing Jupiter's total to fifteen) as well as a ring, like Saturn's but far fainter.

Courtesy of Jet Propulsion Laboratory

This mosaic of Saturn's rings was compiled from photographs taken by NASA's Voyager I *on November 12, 1980. NOVA looked at Voyager's stunning discoveries in "Resolution on Saturn."*

As the two spacecraft rounded the planet, each stole a boost from Jupiter's angular momentum and accelerated toward Saturn. This technique, called gravity assist, is being used with growing sophistication as a substitute for raw rocket power. It permitted the Voyagers to visit more than one destination, and in far less time than conventional trajectories would have allowed. A follow-on mission to Jupiter, called *Galileo*, was launched in 1989 and made use of multiple gravity assists at Venus and Earth to arrive at Jupiter in 1995. In 1990 an ESA probe called *Ulysses* first headed in the wrong direction—to Jupiter—to get a powerful gravity assist into a completely different orbital plane so it could observe the polar regions of the sun. With the growing reliability of spacecraft and the decreasing depredations of the Ghoul, there is no longer such a rush to get probes to their destinations, leaving time for complicated maneuvers to replace heavy, expensive rockets.

At Saturn the Voyagers revealed unimagined complexity in the ring system, which consists of thousands of bands that scientists theorize are shaped by the effects of moons, meteoroids, magnetic fields, and chaotic interactions among particles. The particles themselves may originate from a moon-sized ice object that wandered too close to the planet and was ripped apart by tidal forces. The Voyagers also spent much of their time rubbernecking Saturn's moons, which are no less bizarre than Jupiter's. Titan, the largest, is the only planetary satellite that has an appreciable atmosphere—a sort of nitrogen smog that hides its surface. Temperature and other data suggest that Titan may possess continents of ice and oceans of liquid ethane. In 2004 a joint NASA/ESA mission called *Cassini/Huygens* will arrive in orbit around Saturn and dispatch a probe to reveal more about this odd world.

Voyager 1's close pass by Titan precluded any planetary encounters beyond Saturn, but *Voyager 2*'s trajectory allowed an extended mission to Uranus and Neptune, bonus objectives made possible by a gravity assist—and a budgetary assist. By the time *Voyager 2* arrived at Uranus in 1986, it was operating far beyond its intended lifetime and range. Uranus is twice as far from the sun as Saturn, which means that sunlight is only a quarter as strong and exposure times must be quadrupled. At Neptune, where *Voyager 2* arrived in 1989, available light dims by half again. (The sun at Uranus is about a thousandth as bright as at Earth.) In the faint light, *Voyager 2* showed new rings around Uranus and Neptune, new details about their atmospheres, new moons, and plenty of new mysteries. One of the last places it photographed was Neptune's moon Triton, the coldest place in the solar system, sporting geysers of frosty nitrogen.

Voyager 2's itinerary did not include Pluto, which must await a future explorer. For now, the Voyagers—along with Pioneers *10* and *11*—are adrift in the Milky Way, sailing farther and farther from the solar system. They are the fastest objects ever launched from Earth, yet they will take roughly forty thousand years to cover the average distance between stars. Interstellar space is so empty that these probes will probably never again come within striking distance of a planetary system. In twenty years we will lose contact with the Voyagers as they deplete their radioactive power generators and fall silent. Until then they will continue transmitting information on the particles and fields at the outer limits of the sun's sphere of influence in the great empty ocean of space.

Where Next?

NASA, Russia, ESA, and Japan have lately launched probes to asteroids, comets, and the Moon, and a panoply of missions are in the works.[4] But after spacecraft have exhausted the mysteries of the solar system, where next? The Voyagers and Pioneers show that we have run into a formidable limit: the stars may be forever out of reach to on-the-spot investigation—they are just too far away. Still, that doesn't mean we can't explore them from a distance. The tools we are using right now may be the most remarkable space missions of all, for they allow us to travel through unimaginable realms of space and time.

Telescopes have been around since the early 1600s. Nowadays they are stationed in Earth orbit as well as on the ground, but both types are first-class space-exploring machines that focus light from the most distant regions of the universe. Of course, telescopes in space can do this more comprehensively, since Earth's atmosphere screens out much of the electromagnetic spectrum and also hampers an instrument's resolving—or focusing—ability. There have been many space telescopes, starting with brief forays on suborbital rocket flights in the late 1940s and graduating to the Orbiting Astronomical Observatory series of satellites (first launched in 1968), the Small Astronomical Satellite series (1970), the Astronomical Netherlands Satellite (1975), and a pair of High Energy Astrophysics Observatories (1977 and 1978). All these instruments explored the ultraviolet and x-ray regions of the sky, revealing extremely energetic processes surrounding awe-inspiring objects such as neutron stars, quasars, and black holes, and giving the universe a new reputation as a dynamic if not downright dangerous place to be.

In the low-energy regime, satellites such as the Infrared Astronomical Satellite (1983) and the Infrared Space Observatory launched by ESA (1995) have unveiled cooler cosmic processes occurring inside dense clouds of dust. These events are connected with star and planet formation and are completely invisible in the optical waveband. In the coldest observations of all, the Cosmic Background Explorer (1989) mapped the sky in microwave wavelengths, corresponding to a mere three degrees above absolute zero. Showing irregularities in the fossil radiation left over from the Big Bang, these are the minute fluctuations in matter density that eventually gave rise to galaxies.

The best-known space instrument is the Hubble Space Telescope (HST), the first orbiting observatory that is in a class with the biggest ground-based instruments. Its 2.4-meter (95-inch) mirror puts HST among the top-thirty optical telescopes ever built—and number one in mirror quality and hence resolving power despite its initial blurry vision. After HST was found to have focusing problems soon after launch in 1990, investigation showed that the mirror's curvature was slightly but disastrously awry. Fortunately, correction proved possible, with, in effect, contact lenses: astronauts installed prescription optics for the cameras and other observing instruments during a space shuttle mission in 1993 and HST has seen perfectly ever since.

From its orbit about 380 miles up, HST can capture Jupiter with virtually Voyager-quality images. This ability was put to spectacular use when a string of comets collided with the giant planet in 1994 and there was no time to dispatch a space probe to the scene. In 1997 HST turned its eye on Mars to provide weather updates for *Mars Pathfinder* and *Mars Observer* as they approached the red planet. HST operates as a kind of space probe on call, able to keep tabs on events all around the planetary neighborhood. The orbiting telescope has also given us the sharpest views yet of Pluto and its moon Charon.

HST has similarly taken stunning photographs of stars and planetary systems in the process of forming in nearby nebulae ("nearby" meaning thousands of light-years away). Looking even farther afield, HST has helped pin down the distances to galaxies by resolving stars of known intrinsic brightness. Such stars are "standard candles" that permit galactic range-finding: the dimmer, the farther. Since we also know how fast galaxies are receding from us by measuring their electromagnetic "redshifts," we now have a much more accurate measure of the expansion rate of the universe and hence its age, now estimated at seven to seventeen billion years.

And in a pair of immensely long deep-space expo-

Harvard College Observatory

NOVA charted the worldwide preparations for the 1986 arrival of one of the solar system's most famous celestial bodies in "Halley's Comet."

sures made in 1996 and 1998, HST photographed practically all there is to see in two seemingly empty spots of sky near the north and south celestial poles, where thousands of galaxies turned up, including some of the faintest ever seen. A few appear as they were more than ten billion years ago (the farther away something is, the longer its light takes to reach us, so we see it in the past). Just as our robotic probes allow us to investigate the milieu in which Earth evolved, instruments such as HST reveal scenarios like those that may have engendered our solar system and even our galaxy: nascent galaxies in collision, supernovae enriching the cosmos with the precursor elements of life, protoplanetary disks forming from immense clouds of dust and gas... Looking at such pictures is a little like reading a novel about someone else's eventful life.

Other space telescopes will continue this voyeuristic work by investigating the most persistent mysteries of the cosmos. The Advanced X-ray Astrophysics Facility (1999) will probe deeper into the enigma of black holes and other energetic objects. The Space Infrared Telescope Facility (2001) will provide our best views yet of the conditions in which planets form and, through spectroscopic measurements, may be able to detect evidence of carbon and water—harbingers of life. The Next Generation Space Telescope (2007) will gather nine times more light than HST and image the as-yet-unseen epoch when galaxies first formed in the universe. The Terrestrial Planet Finder, still a wishful gleam in NASA's eye, may take off in about fifteen years on a mission to locate Earth-like planets, if there are any, around the closest few hundred stars. Should such planets turn up, a truly humongous telescope will doubtless be launched to obtain pictures reminiscent of the whole-Earth images shot by astronauts on the Moon. Such awesome images could change the course of human thought.

For humans the next steps in space are less clear. The first space shuttle in 1981 inaugurated a piloted transportation system still in use, but the shuttle has proved much more expensive to operate per pound of payload than traditional expendable rockets. The reason, of course, is that the people aboard require a considerably safer and thus more complex vehicle. The *Challenger* accident in 1986, which killed seven crew members on a routine satellite-launching mission, made it tragically obvious that there is no reason to send people to perform jobs easily done by machines.

If the shuttle has been a failure as a low-cost launcher, it has proved to be an interesting place to do research, though much of this work is geared toward learning how humans can live in space rather than toward exploring the space environment. The international space station in Earth orbit, now under construction, will similarly test the technology needed to keep people healthy for years at a time—on Mars missions, for example. Space station crew members will also investigate new technologies that take advantage of weightlessness, such as producing ultrapure crystals for high-tech applications and growing tissue samples for medical research, though the payoff from such efforts is unlikely to justify the enormous cost of the space station, which is on par with the Apollo project.

In sum, human spaceflight encompasses something besides scientific rewards. The Apollo voyages to the Moon were not really about lunar science, nor did they pursue anything concrete such as gold and other riches and expanded empire—the objectives that impelled Columbus. The international space station is more of a symbolic adventure than a quest for new knowledge. One day we will go to Mars for the same reasons—or lack of them—even though robots could easily do everything we will.

But in a sense our space missions *have* struck gold: pictures. Through them we have participated in history's grandest adventure, witnessing through the eyes of our machines views that astronomers speculated about for centuries: the far side of the Moon, the surface of Mars, Earth as a pale blue dot from the edge of the solar system. Instruments such as HST have shown us scenes that stretch the bounds of comprehension: a pillar of gas a light-year long, shedding stars surrounded by dusty cocoons each the size of our solar system. Space flight may always be limited to a few intrepid adventurers, but space exploration is available to all—through information transmitted from our wide-roaming robots and far-seeing telescopes. Anyone can journey to Mars or Io or Saturn, the Eagle Nebula or the Andromeda Galaxy. All that's required is a stack of photos, an inquiring mind, and a vivid imagination. It's an incredible trip.

NOTES:

1. *Salyut* evolved through successive models into the venerable *Mir*, in orbit since 1986.
2. In 1996 scientists announced that one of the meteorites showed signs that microbial life once existed on Mars. Serious doubts have since been raised about this evidence.

The case for life on Mars—or anywhere else beyond Earth—is still unsubstantiated.

3. Russia inherited the space-exploring infrastructure of the former Soviet Union, while Japan and ESA, a consortium of western European nations, have been actively involved in space research since the 1980s.
4. Future missions include *Mars Surveyor Lander* (1999), a NASA spacecraft targeting the intriguing polar region of the red planet will launch two microprobes to penetrate the martian surface; *Stardust* (1999), a NASA rendezvous with a comet designed to return samples of interstellar dust; *Nereus Sample Return* (2002), a proposed Japanese mission to an asteroid; *Rosetta* (2003), an ESA asteroid flyby and comet rendezvous, including a science station to be deployed on the comet's nucleus; and *Mars Sample Return* and *Phobos Sample Return* (2005), proposed joint American-Russian efforts. There is also talk of returning to Europa and sending an inaugural probe to Pluto.

SCIENTIST'S NOTE:

The Future of Flight

Tom D. Crouch

Senior Curator, National Air and Space Museum

"No airship will ever fly from New York to Paris," Wilbur Wright remarked to an Illinois reporter in 1909. "That seems to me to be impossible." The engine was the problem. "No known motor can run at the requisite speed for four days without stopping," he explained, "and you can't be sure of finding the proper winds for soaring." Nor did he hold out any great hope for improved carrying capacity. "The airship will always be a special messenger," he predicted, "never a load-carrier."

Of course, Wilbur Wright's limited vision of the future for the technology that he and his brother had pioneered seems incredibly shortsighted. But the inability of a brilliant inventor to forecast even the near-term progress of an infant technology is not so difficult to understand. In 1909, the airplane had nowhere to go but up, and much faster at that. Aeronautical progress was occurring at such a rapid rate that the world's first military airplane, the aircraft that the Wright brothers sold

In 1983, astronaut Sally Ride became the first American woman launched into space. In "Space Women," NOVA looked at the experiences of America's female astronauts.

to the U.S. Army in 1909, was judged obsolete and presented to the Smithsonian as a historical object less than two years after it entered service. In such a period, predicting the direction in which technology will move over the next few decades can be extremely tricky.

The aeronautical prognosticator of today has a much easier time of it. We are approaching the end of a century of progress during which machines conceived and crafted by human minds and hands have flown from the sands of Kitty Hawk to a point beyond the edge of the solar system. The high-performance products of a mature aerospace industry are safer, more reliable, and much longer-lived than the pioneer aircraft of Wilbur Wright's day. Boeing 777 aircraft now entering service will still be flying three decades from now. Even more astonishing, *Aviation Week and Space Technology* recently reported that the B-52, B-1, and B-2 bombers that the U.S. Air Force now operates will still be flying in the year 2030. If true, the B-52 will have remained in active service for almost eighty years! What automobile, railroad locomotive, or naval vessel can match such a record?

That being the case, it is no trick at all to comment on the probable state of flight technology a quarter century from now. The airplanes that will operate then are either in the air now or evolving on the personal-computer screens of engineers from Seattle to Toulouse. Obviously, commercial aircraft could fly faster and higher, but it is by no means certain that the resulting profits would justify the effort. The new benchmarks of aeronautical progress call for us to fly cleaner, quieter, safer, and more economically. Efficiency, both aerodynamic and economic, has emerged as a force that will drive and direct aeronautical change in the decades to come. But then, I sound just as conservative as Wilbur Wright.

BACK TO THE FUTURE

Putting a Human Face on Science: NOVA Turns 25

Susan Reed

Science Journalist
Senior Editor, WGBH

Since 1974, *NOVA* has demystified the wonders of science, taking a nation of viewers from the depths of the ocean floor to the eye of a hurricane, from the rim of a simmering volcano to the beginning of life itself deep inside the womb. *NOVA* has become such a staple of American television—casting what *The New York Times* recently called "one of the steadiest lights in a firmament of more ephemeral fare"—that few remember the cool skepticism among broadcasters and critics when *NOVA* first made its debut.

"The common wisdom was that science wasn't suitable for television," says *NOVA* executive producer Paula Apsell, who has led the award-winning series since 1984. "WGBH and PBS [the Public Broadcasting Service] thought otherwise, and launched an ambitious experiment. During the next quarter century that experiment, *NOVA*, has irrefutably proved that Americans have not only an interest in but an insatiable appetite for science stories."

A Star Is Born

In May 1971, former WGBH producer Michael Ambrosino sat at his desk in London, where he was pursuing a yearlong fellowship at the BBC (the British Broadcasting Corporation), and wrote a six-page letter to WGBH vice-president for programs Michael Rice outlining a new public television series. Modeled after the BBC's science documentary series *Horizon*, the new program would "examine how the world worked, using the scientific process of discovery as a narrative device to tell good stories." WGBH liked the idea and Ambrosino returned to Boston to become executive producer, assembling three production teams headed by veteran British filmmakers of science documentaries, in preparation for the series' launch in 1974.

But what would the series be called? As the air date approached, the list of possible titles grew longer; among the frontrunners was *Eureka*. "To me, *Eureka* stood for everything the new series wasn't: science where solutions to the world's problems are created instantly, out of pure inspiration," Ambrosino recalls. "Our series was about scientists paying diligent attention to detail, sharing information with others, demanding proof." In the end, Ambrosino chose NOVA because "it represented something big, bright, new, and bold, something to which you had to pay attention."

Great Stories of Science

"Science films are the same as any other high-quality television documentaries, in that they have to have a good story with a beginning, middle, and end, interesting people, and beautiful visuals," Apsell says. "At the same time, we're often dealing with content that is difficult to understand and scientists who sometimes need to be coached to express their ideas in a way that a general audience can follow. Creating such programs is a special art."

What lies at the heart of *NOVA*'s enduring appeal? "Putting a human face on science," Apsell responds. "We want viewers to get to know scientists as human beings, to see the passion and commitment they bring to their work. We've also increasingly emphasized the adventure of science—not simply the intellectual adventure but the physical adventure. As technology has developed, it's given scientists a greater ability to do on-site research, whether at the summit of Everest or in the eye of a storm. We show viewers science as it happens, and take them places they might not otherwise see."

The series quickly found, and has maintained, an enthusiastic audience (nine million Americans tune in every week) by introducing provocative subjects mirroring the extraordinary growth in many scientific fields during the past quarter century: genetics, anthropology, environmental studies, neuroscience, computers, archeology, geology, and medicine, to name a few. "People hate to be taught but love to learn," Ambrosino says. For twenty-five years, *NOVA* has given viewers ample opportunity to do the latter, providing insight on everything from the science behind Hollywood's dazzling special effects *(Special Effects: Titanic and Beyond)* to Princeton mathematician Andrew Wiles's successful quest to solve history's most famous math problem: Fermat's Last Theorem *(The Proof)*.

Like science itself, the end of one *NOVA* story often leads to the beginning of another. "We've returned to cover subjects time and again as fields develop and change, often in unpredictable ways," says Apsell, who joined *NOVA* as a producer soon after the series' debut. "Back in the 1970s I worked on a film about computers called *The Mind Machines*. At that point everyone who was anyone in computer science believed that the Holy Grail was artificial intelligence: to make a machine that could think like a person, see like a person, use language like a person. Everyone agreed that the future of computers lay in huge, complicated machines that could perform those feats. No one foresaw that this was not the way computers were going to touch people's lives—that within the next fifteen years people were going to have computers of their own, and that computing was going to become decentralized, more personal, and indispensable." *NOVA* returned to the subject in the early '90s with a five-part miniseries called *The Machine That Changed the World.* "One of our greatest strengths is that the subject matter we cover constantly evolves."

Promoting Science Literacy

"Twenty-five years ago, I think the public, and scientists, regarded much of science as the bailiwick of experts," Apsell says. "That's no longer true. Today we live in a world where science and technology are intimately connected with people's lives. Every time people turn on a computer, make a difficult health care decision, or cast a vote on an important environmental issue, their understanding of science and technology informs their actions. Promoting science literacy, helping people navigate through an increasingly complex technologically based world, is *NOVA*'s primary mission."

The opportunities for presenting science's adventure stories have never been greater, nor more important. A 1994 National Science Foundation study reports that television is the public's primary source of information on science. Americans watch an average of forty-two science programs per year, and those with access to cable or direct satellite watch twice that amount, according to a 1996 National Science Foundation report. *NOVA*'s success has spawned a mini-industry, with niche cable channels as well as the commercial networks scrambling to cash in on Americans' interest in science.

"As science programs on cable and network TV proliferate," Apsell says, "we in public broadcasting are proud that not only did we do it first but we still, by far, do it best." The growing presence of science on television is a cause for both celebration and concern, she believes. "Nearly all science programs on cable channels and the commercial networks [as opposed to public television] are produced by entertainment divisions. Many of those programs exploit people's fears and fascination with paranormal phenomena under the guise of science. NOVA provides a valuable alternative to this burgeoning New Age supermarket of pseudoscientific ideas. NOVA counters this trend, not only by rigorously exploring science and technology but through programs that provide rational explanations for sensationalized subjects—like our *Secrets of the Psychics* and *Kidnapped by UFOs?*"

Blueprint for the Future

NOVA has won every major broadcasting award, critical acclaim, and millions of weekly viewers. Scientists regularly laud the series for its portrayal of the process of science and the people who perform it. In May 1998, the National Science Board, the governing body of the National Science Foundation, awarded its first annual Public Service Award to NOVA for "its contributions to public understanding of science and engineering." "*NOVA* set the standard for showing how science is done and what drives those who do it," said Richard Zare, the board's chair. And David Perlman, chair of the award's selection committee and science editor for the *San Francisco Chronicle*, said, "*NOVA* has become an American institution, regularly enthralling its huge PBS audience with clear, accurate, and wide-ranging programs exploring virtually every aspect of science."

With so much accomplished, what will America's most popular and esteemed science series do for an encore? "There's a saying in this business that 'you're only as good as your last program,' and I believe that," Apsell notes. "When you're in the position of running the *NOVA* series, you know you have a national treasure in your hands, one created by some of the most gifted science producers in the world today, many of whom are homegrown talent. *NOVA* was a pioneer in putting science on television, and we need to continue to play that role, producing innovative programs that combine rich content with an entertaining format.

"We're using our twenty-fifth anniversary to look to the future, not the past," Apsell continues. "This is an exciting time to be a science journalist. New technologies are transforming not only where and how science is conducted, but the way *NOVA* can present its stories." One of the most powerful communication tools to emerge in the last few years is the Internet, and *NOVA* has led the way in exploiting the new medium. "We decided it was very important for the foremost science and technology program on American television to have an early and strong presence on the Web," Apsell says. NOVA/PBS Online Adventures (www.pbs.org/nova) debuted with *Everest Quest*, which tracked the ascent of an elite climbing team led by IMAX filmmaker David Breashears in the spring of 1996. The Web site drew worldwide attention when tragedy hit other climbing teams: a lethal storm struck the crowded final route to the summit, leaving eight climbers dead. NOVA Online was there, reporting news of the tragedy, and the science of high-altitude weather, to a worldwide audience.

The following year a NOVA crew returned to Everest to answer questions about the effects of high altitude on climbers' mental and physical abilities, resulting in another NOVA first: the simultaneous production of a live Web site and a NOVA film. "The Internet offers us an extraordinary opportunity to amplify NOVA's educational impact," says Apsell, noting that NOVA's teachers' guides and videos are used by millions of educators and students annually. "You'll be seeing more simultaneous NOVA Web and film productions in partnership with PBS in the years ahead."

Viewers can also look forward to even more beautiful programs on both the small and giant screen, as NOVA makes the transition to digital television and expands its NOVA Large Format Film efforts (70mm films made for

IMAX/IMAXDome theaters around the world). "All of our new television programs are being filmed ready for widescreen digital TV, including the sequel to our most popular program of all time, *The Miracle of Life*. The new film, *Life's Greatest Miracle*, will provide viewers with unprecedented images of human development from conception onward," Apsell says. And building on the success of *To the Limit* (about the physiology of world-class athletes), *Stormchasers*, and *Special Effects*, which was nominated for an Academy Award, NOVA Large Format Films has several more productions in the works, including *Island of the Sharks* and *Volcano: Lost City of Pompeii.*

Will NOVA's well ever run dry? "Scientists are smart, curious people and the questions they pursue are intrinsically fascinating because they help us make sense of the world around us," Apsell responds. "What could be more interesting, or enduring, than that?"

Scientist's Note:

Toward the Ultimate Synthesis

Edward O. Wilson

Pellegrino University Research Professor and Honorary Curator in Entomology, Harvard University

Perhaps it is the imminence of the new millennium that colors my thinking, but I believe we have begun a major intellectual synthesis that will culminate during the next quarter century. This synthesis will see the conjunction of the great branches of learning—the natural sciences, the social sciences, and the humanities—which, four centuries after Francis Bacon launched the dream in *The Advancement of Learning* and other writings, will soon be proved to share a common foundation.

I call this interlocking of testable cause-and-effect explanations across disciplines *consilience*. During the past quarter century the natural sciences have attained a remarkable degree of consilience, moving from one level of complexity to the next—from physics to chemistry to molecular and cellular biology, and from there to human behavioral genetics, the brain sciences, and evolutionary biology. These last subjects in turn are expanding outward from the traditional natural sciences toward the social sciences and humanities, illuminating a borderland of little-known phenomena that connect brain to mind and mind to culture. Meanwhile, reaching from the social sciences toward the natural sciences are disciplines such as cognitive psychology and biological anthropology. Consilience across all the great branches of learning is still unproven but—in my opinion and that of a growing number of researchers—a very likely prospect.

Where this unity will lead in the common decades is mostly unpredictable. But I have no doubt it will open new domains of inquiry and challenge scholars of the great branches to the utmost. At the very least it will lay a foundation beneath the social sciences by providing a more exact knowledge of human nature, which today can be better defined as the inherited regularities of mental development. That knowledge in turn will permit the long-sought axiomatic theory from which social scientists can build more reliably predictive models. In the humanities, foundational unity will similarly deepen our understanding of the nature of aesthetics and of moral reasoning.

I believe further that a more consilient approach is the necessary means to revitalize the liberal arts, which are flagging badly. By examining the cause-and-effect connections among the natural sciences, social sciences, and humanities, students can better understand each of these domains of learning. Practitioners will also have more effective tools with which to examine the eternal questions of reflective thought, which presumably are still the focus of the liberal arts: Who are we, where do we come from, how shall we decide where to go?

APPENDICES

About NOVA

Since it first aired in 1974, *NOVA* has defined science television for the US and the world. *NOVA* is watched by an average of 8.5 million people each week. *NOVA* films are also seen by viewers in 52 other countries, from Australia to Zimbabwe.

The series has been honored with virtually every major broadcast industry award, including the Emmy, the Alfred duPont–Columbia University Award and the George Foster Peabody Award.

NOVA helps its legions of fans—men, women and children of all ages—explore the science behind the headlines, along the way demystifying science and technology . . . and the people who make science happen.

NOVA is now celebrating its 25th anniversary season of quality science programming on PBS. The season began with a celebratory party at the United States Naval Observatory on September 16, highlighting the first broadcast of the season—*Lost at Sea: The Search for Longitude*, based on author Dava Sobel's best-selling book *Longitude*—with Vice President Al Gore as the featured speaker.

Critical Acclaim for *NOVA*:

"...probably the most consistently intelligent program on national television." —*Washington Post*

"A remarkable series that has no equal on television." —*New York Daily News*

"Public television's gold medalist of a science show... *NOVA*." —*Chicago Tribune*

"Consistently intelligent, provocative, and well produced." —*Newsday*

"Television's most consistently excellent science program." —*The Christian Science Monitor*

"Anything that *NOVA* tells you, automatically believe. It's that wise and dependable." —*Satellite Direct*

"News minus sensation: Sober without being stuffy, provocative without being crude, these programs endure without the distractions of hype or faddish gimmicks. Dramatic, clinical, inspiring, challenging and ultimately euphoric, this *NOVA [Coma]* works on every level." —*USA Today*

"I can say, without qualification, that this is the best documentary program on mathematics I have ever seen *[The Proof]*." —*New York Times*

NOVA ADVISORS

Dr. Harvey V. Feinberg
Provost
Harvard School of Public Health

Dr. Daniel Fox
President
Milbank Memorial Fund

Dr. Robert Hazen
Clarence Robinson Professor of Earth Science
Carnegie Institution of Washington and George Mason University

Dr. Shirley Ann Jackson
Chairwoman
United States Nuclear Regulatory Commission

Dr. Harry Woolf
Professor at Large Emeritus
Princeton University Institute for Advanced Study

Mr. George Strait
ABC News

Dr. Philip Morrison
Institute Professor and Professor of Physics
Massachusetts Institute of Technology

Ms. Phyllis Morrison
Teacher, Author

Dr. Arthur Kantrowitz
Professor
Dartmouth College

Dr. Steven J. Marcus
Science and Medicine Editor
Minneapolis Star Tribune

NOVA STAFF

Paula Apsell, Executive Producer
Alan Ritsko, Managing Director
Melanie Wallace, Senior Producer for Coproductions and Acquisitions
Evan Hadingham, Senior Science Editor
Stephen Lyons, Senior Editor for Program Development

Michael Barnes, Producer
Dick Bartlett, Editor
Mary Brockmyre, Associate Producer
Carl Charlson, Producer
Kate Churchill, Associate Producer
Liesl Clark, Producer
Julia Cort, Producer
Andrea Cross, Production Assistant
Deborah Fryer, Associate Producer
Albert Lee, Research Assistant
Susan Lewis, Producer
Nancy Linde, Producer
Marti Louw, Associate Producer
Joe McMaster, Producer
Jackie Mow, Associate Producer
Stephanie Munroe, Editor
Stephen Sweigart, Associate Producer
Caroline Toth, Associate Producer

Susanne Simpson, Executive Producer, NOVA Large Format Films
Kelly Tyler, Associate Producer, NOVA Large Format Films
Lisa Roberts, NOVA Large Format Films

Lauren Aguirre, Senior Online Producer
Peter Tyson, Online Producer
Kim Ducharme, Online Designer
Brenden Kootsey, Technologist
Rob Meyer, Online Production Assistant

Karen Hartley, Project Director,
Teachers' Guides

Lisa Mirowitz, Coordinating Producer/Science Unit

Mark Geffen, Post Production Supervisor
Rebecca Nieto, Post Production Editor
Carla Fremlin, Associate Producer, Post Production
Franziska Blome, Post Production Assistant

Paul Marotta, Senior Publicist
Thalassa Skinner, Publicity
Diane Buxton, Publicity

Laurie Cahalane, Business Manager
Nancy Marshall, Paralegal
Jessica Maher, Unit Manager
Cesar Cabral, Unit Manager
Queene Coyne, Production Secretary
Linda Callahan, Production Secretary

Major funding for *NOVA* in its twenty-fifth anniversary season is provided by the Park Foundation, The Northwestern Mutual Life Foundation, and Iomega Corporation. Additional funding is provided by the Corporation for Public Broadcasting and public television viewers. *NOVA* is a production of WGBH Boston.

NOVA AWARDS

1998

The New York Festivals
Silver World Medal
"Titanic's Lost Sister"
Bronze World Medal
"Secrets of Lost Empires:Obelisk/Inca"

National Headliner Awards
First Place—Documentary or Series
"Coma"
Best Of Show—Television
"Coma"

National Science Board
Public Service Award
NOVA

American Association for the Advancement of Science
Science Journalism Award
"Warnings From the Ice"

1997

Chicago International Television Festival
Gold Plaque
"Plague Fighters"
Certificate Of Merit
"Einstein Revealed"

George Foster Peabody Award

Odyssey of Life: Parts 1–3

Part 1—The Ultimate Journey

Part 2—The Unknown World

Part 3—The Photographer's Secrets

Society For Technical Communication

President's Award

NOVA

1996

National Education Association Award for Advancement of Learning Through Broadcasting

NOVA—The Series

News/Documentary Emmy Award

Outstanding Background/Analysis of a Single Current Story (Programs)

"Siamese Twins"—WGBH Science Unit

Festival International du Film Scientifique du Quebec

"Lightning!"

Banff Television Festival

"B-29 Frozen in Time"

"Plague Fighters—Ebola, Inside an Outbreak"

Alfred I. Dupont Columbia University Award

Silver Baton

"Plague Fighters"

1995

Carl Sagan Award

Outstanding Science Television Series

NOVA and the WGBH Science Unit

News/Documentary Emmy Award

Outstanding Informational/Cultural Program

"Secret of the Wild Child"—WGBH Science Unit

1994

Ohio State Award

"Wanted: Butch and Sundance"

Genesis Award

Outstanding PBS Documentary

"The Great Wildlife Heist"—WGBH Science Unit

1993

National Association of Black Journalists:

Outstanding Coverage of the Black Condition

presented to George Strait and Denise DiIanni for "Deadly Deception"

Ohio State Award

"The Machine That Changed the World"—WGBH Science Unit

National Headliner Award

"Iceman"

1992

American Heart Association
Howard W. Blakeslee Award
"Avoiding the Surgeon's Knife"

Festival International des Images du Ciel, de l'Espace et de l'Homme
Prix Du Reportage
"The Russian Right Stuff: The Mission"

News/Documentary Emmy Award:
Outstanding Background Analysis of a Single Current Story (Programs)
"Suicide Mission to Chernobyl"

News/Documentary Emmy Award:
Outstanding Historical Programming (Programs)
"Russian Right Stuff"

American Association for the Advancement of Science
Westinghouse Science Journalism Award
"Eclipse of the Century"

Genesis Award
"Saddam's War on Wildlife"

George Foster Peabody Award
"Machine that Changed the World"

1991

Chicago International Film Festival
Silver Plaque Award
"Russian Right Stuff"

Columbus International Film And Video Festival

"Bomb's Lethal Legacy"

1990

French Ministries Of Technology And Communication

Prix Jules Verne, Finalist

WGBH Science Unit and NOVA for its contribution to science broadcasting

American Association for the Advancement of Science

Westinghouse Science Journalism Award

"Hurricane!"

Ohio State Awards

"Can the Vatican Save the Sistine Chapel?"

1989

Ohio State Award

"Buried in Ice"

"Can the Vatican Save the Sistine Chapel?"

NATAS News And Documentary Emmy Award:

"Race for the Superconductor"

"Can the Vatican Save the Sistine Chapel?"

BBC Award For Best Science Film 1987–1989

"Death of a Star"

British GAS Award

NOVA and Horizon (BBC) for making the best science films in the world

1988

Ohio State Awards

"Why Planes Crash"

"Can AIDS Be Stopped"

George Foster Peabody Award

"Spy Machines"

American Film Festival Blue Ribbon

"A Man, a Plan, a Canal, Panama"

1986

Ohio State Award

Natural And Physical Sciences

"Acid Rain: New Bad News"

International Agricultural Film and Television Competition Award

"Farmers of the Sea"

American Psychology Association National Media Award

"Life's First Feelings"

American Institute of Aeronautics and Astronautics

New England Section Media Award

"The Plane That Changed the World"

National Educational Film and Video Festival Award

"Tornado!"

"What Einstein Never Knew"

San Francisco International Film Festival
Golden Gate Award
"Seeds of Tomorrow"
Distinguished Participation Award
"What Einstein Never Knew"

Aviation/Space Writers Association (AWA)
Journalism Award
"The Planet That Got Knocked On Its Side"

Chicago International Film Festival
Gold Hugo Award
NOVA—The Series

Earthwatch Film Awards
"Return of the Osprey"
"Stephen Jay Gould: This View of Life"

1985

San Francisco International Film Festival
Golden Gate Award
Best Network Nature/Science Program
"Acid Rain: New Bad News"

NATAS News and Documentary Emmy Awards
Outstanding Background/Analysis Of A Single Current Story
"Acid Rain: New Bad News"
Outstanding Coverage Of A Continuing News Story
"AIDS: Chapter One"

1984

San Francisco International Film Festival
Special Jury Award
"The Miracle of Life"

Ohio State Award
"The Miracle of Life"

American Film Festival
Blue Ribbon: Science/Nature
"The Miracle of Life"
Red Ribbon: Environmental Issues
"Asbestos: A Lethal Legacy"
Red Ribbon: Health/Fitness
"Fat Chance in a Thin World"
Emily Award: Program which received the highest score of all programs entered in the festival
"The Miracle of Life"

Parents' Choice Award
Ongoing and Continuing Excellence
NOVA—The Series

International Film and Television Festival of New York
Gold Medal: Documentary Series
NOVA—The Series

1983

Ohio State Awards:
Natural and Physical Sciences
NOVA—The Series
Social Sciences and Public Affairs
NOVA—The Series

National Association to Aid Fat Americans
"Fat Chance In a Thin World"

Robert F. Kennedy Journalism Award
"Finding a Voice"

San Francisco Film Festival
Certificates of Participation
"Adventures of Teenage Scientists"
"Test Tube Babies: A Daughter for Judy"
"Animal Imposters"
"Artists in the Lab"
"Goodbye Louisiana"
"Life: Patent Pending"
"Notes of a Biology Watcher"
"Why America Burns"

Southern California Motion Picture Council for Outstanding Family Entertainment
Golden Halo Award
NOVA—The Series

International Film Festival of New York
NOVA—The Series

Audubon Festival Award
Conservation/Ecology
"Fragile Mountain"
Land And Water Resources
"Goodbye Louisiana"

Audubon Certificate of Excellence
Conservation/Ecology
"Salmon on the Run"
Nature/Wildlife
"A Field Guide to Roger Tory Peterson"

"Animal Imposters"

International Wildlife Film Festival

"Salmon on the Run"

"Animal Imposters"

American Film Festival

Blue Ribbons

"Tracking the Supertrains"

"Life: Patent Pending"

NATAS News And Documentary Emmy Award

Special Classification Of Outstanding Program Achievement

"The Miracle of Life"

Massachusetts Medical Association Award

"Aging: The Methuselah Syndrome"

George Foster Peabody Award

"The Miracle of Life"

1982

NATAS News and Documentary Emmy Award

"Here's Looking At You, Kid"

American Association for the Advancement of Science

Westinghouse Science Journalism Award

"Anatomy of a Volcano"

Ohio State Award

Natural And Physical Sciences

NOVA—The Series

International Wildlife Film Festival
"Notes of a Biology Watcher"

Audubon International Film Festival Award
"The Sea Behind the Dunes"

Cine Golden Eagle Awards
"Palace of Delights"
"The Television Explosion"

Wildscreen Festival Award
"The Sea Behind the Dunes"

American Film Festival
Blue Ribbons
"Cancer Detectives of Lin Xian"
"Anatomy of a Volcano"
"Salmon on the Run"
Red Ribbons
"Moving Still"
"Notes of a Biology Watcher"

International Rehabilitation Film Festival Award
"Finding a Voice"

International Film and Television Festival of New York
"Palace of Delights"

1981

Ohio State Award
"Life On a Silken Thread"

American Psychological Association
"The Pinks and the Blues"

National Audubon Society Award
"Black Tide"

Prix Futura
"The Doctors of Nigeria"

NATAS News and Documentary Emmy Award
"Why America Burns"

American Water Resources Association
"The Water Crisis"

American Film Festival
Red Ribbon
"The Sea Behind the Dunes"

Chicago Film Festival Award
NOVA—The Series
"The Great Violin Mystery"

1980

National Wildlife Federation Award
NOVA—The Series

Alfred I. Dupont-Columbia University Award
"A Plague On Our Children"

1978

American Film Festival
Blue Ribbon
"The Green Machine"
Red Ribbons
"The Gene Engineers"

"Across the Silence Barrier"
"Dawn of the Solar Age"
"The Sunspot Mystery"

Alfred I. Dupont-Columbia University Award
NOVA—The Series

1977

Alfred I. Dupont-Columbia University Award
"Incident At Brown's Ferry"

American Film Festival
Red Ribbons
"A Desert Place"
"Why Do Birds Sing"
"Strange Sleep"
"Where Did the Colorado Go?"

Ohio State Awards
NOVA—The Series
"The Business of Extinction"

1975

George Foster Peabody Award
NOVA—The Series

Prix Italia
"Plutonium Connection"

1974

American Psychological Association
"The First Signs of Washoe"

NOVA Programs

Perfect Pearl, The 2520
December 29, 1998
Producer(s): Stephen Sweigart, NOVA; Megan McMurchy, Film Australia

Pearls are back in fashion—in a big way. For thousands of years, humans have speculated about the mystery of the pearl, a unique gem produced by a living animal that requires no mining, extraction, cutting, or polishing to reveal its beauty. *NOVA* travels around the world to exotic locations where rare pearls are harvested by divers, and to farms where huge numbers of pearls are grown. Will the cultured pearls ruin the value of those grown in the wild? From the depths of the ocean to the riches of Fifth Avenue, *NOVA* revels in the luster of these desirable gems.

Leopards of the Night with David Attenborough 2519
December 1, 1998
Producer(s): Stephen Sweigart, NOVA; Amanda Barrett & Owen Newman, BBC

Night stalkers by nature, leopards are observed both night and day, using state-of-the-art camera equipment, to reveal that leopard society is much more complex and far less solitary than originally believed. Filmed in the Luangua Valley in Zambia, this film reveals the challenges and dangers faced daily by these beautiful animals. Shadowed by hungry hyenas in pursuit of leftovers, and stalked by lumbering crocodiles hoping to tackle a lone leopard on a kill—how can they hope to challenge such beasts?

Ice Mummies—Frozen in Heaven 2516
November 24, 1998
Producer(s): Julia Cort, NOVA; Tim Haines, BBC

The first of a three-part series on the science of the frozen past. The bizarre and fascinating story of the remains of Inca culture, frozen for posterity high in the mountains of the Andes. Evidence has emerged of sacrifice to the mountain gods, whose existence dominated the civilization more than 500 years ago. The film traces the frozen bodies of children uncovered by archeologists in South America, and follows an archaeological expedition to a high-altitude sacred site in search of ritual remains and another body. How did they come to be there? Why did they go to their deaths willingly? What was the religious framework that dictated their sacrifice to fierce gods?

Ice Mummies—Return of the Iceman 2518
November 24, 1998
Producer(s): Joe McMaster, NOVA; Tim Haines, BBC
Cutting-edge science and archaeology are reconstructing the life and culture of The Iceman—the 5,000-year old frozen corpse found buried in the ice of the Alps. By analyzing every inch of the Iceman's body and the tools and equipment found with it, scientists are piecing together the most complete picture yet of the late Stone Age in this part of Europe. X-rays, CAT scans, and microscopic analysis of this spectacular find are revealing where the Iceman lived, what he ate, and how he may have died. Nuclear physics reveals that the Iceman's hair was contaminated with arsenic and copper, suggesting he was involved in copper production centuries before it was known to exist in the region.

Ice Mummies—Siberian Ice Maiden 2517
November 24, 1998
Producer(s): Susan Kopman Lewis, NOVA; Andrew Thompson, BBC
The Siberian Ice Lady, discovered in the Pastures of Heaven, on the high Steppes, is believed to have been a shamaness of the lost Pazyryk culture. She had been mummified and then frozen by freak climactic conditions around 2,400 years ago, along with six decorated horses and a symbolic meal for her last journey. Her body was covered with vivid blue tattoos of mythical animal figures. Together with the newly discovered body of an ancient man nicknamed

"Conan," her body has now been restored and is providing new clues as to the role and power of women in the nomadic peoples of ancient Siberia.

Deadly Shadow of Vesuvius 2515
November 10, 1998
Producer(s): Denise DiIanni, NOVA
At present there are more than 1,400 active volcanoes on this planet. Many of these are either at sea or in remote places where they present little risk to anyone. A small number are situated in the middle of populated areas and, although inactive, have troublesome histories. One is Mt. Rainier, which overlooks Seattle, and the other is Mt. Vesuvius, which dominates the Bay of Naples. One does not have to look far beyond the shattered remnants of the Roman city of Pompeii to understand the risk that Vesuvius presents today. The volcano has remained dormant since 1944, but geological evidence suggests that Vesuvius is on the move again. The neighboring city of Pozzouli is being torn apart. More than two million people live under the shadows of Vesuvius. This program looks at new scientific measurements of this infamous volcano, at the threat posed by a new eruption, and at the historical day of August 24, 79 A.D. when Pompeii died.

Special Effects Titanic and Beyond 2514
November 3, 1998
Producer(s): Bill Lattanzi, Susanne Simpson & Kelly Tyler, NOVA
NOVA goes behind the scenes of Hollywood's biggest blockbuster ever, *Titanic*, and lifts the curtain on how James Cameron achieved his spectacular vision. The secrets behind the explosions on the set of *The X-Files Movie* and the painstaking work that went into having a computer-animated Flubber are revealed. Although the computer has made such effects both easier to produce and more likely to fool the eye, there is nothing new about them. In this star-studded NOVA, the art of illusion meets the science of perception.

Terror in Space 2513
October 27, 1998
Producer(s): Joe McMaster, NOVA; Jill Fullerton-Smith, BBC
Harrowing and life-threatening problems aboard the aging Mir space station are seen through the eyes of the Russian and American astronauts who lived through them. Feel the heat from the fire that erupted on board. See the collision between Mir and another space craft. Endure the power outages and the computer failures that have jeopardized lives. Hear the debate over whether NASA should continue to risk the lives of its astronauts by sending them to Mir in preparation for the launch later this year of the most ambitious space project yet—the International Space Station.

Chasing El Nino! 2512
October 13, 1998
Producer(s): Carol Fleischer, NOVA
A massive planet-sized machine controls our weather day-to-day, and our climate season-to-season. It takes an event of staggering proportions to disrupt a machine this large and powerful, a juggernaut with more energy than a million nuclear bombs. Signs now indicate that such an event is underway—El Nino. More than a series of storms stunning the California coastline, El Nino is second only to the seasons in its effect on global weather. In a P-3 airplane off the California coast, a team plunges into a storm front to explore its cause and effects. In a boat off the Galapagos, an array of buoys are checked for temperature and current data. On a mountain in Peru, signs of the devastation of past El Ninos are revealed. As scientists push to extremes to explore this phenomenon, they understand for the first time the extent to which all the world's weather is connected, and just how delicate is the balance.

Lost At Sea: The Search For Longitude 2511
October 6, 1998
Producer(s): David Axelrod & Peter Jones, NOVA
It was one of humankind's most epic quests—a technical problem so

complex that it challenged the best minds of its time, a problem so important that the nation that solved it would rule the economy of the world. The problem was navigation by sea—how to know where you were when you sailed beyond the sight of land—establishing your longitude. While the gentry of the 18th Century looked to the stars for the answer, an English clockmaker, John Harrison, alone for decades to solve the problem. His elegant solution made him an unlikely hero and remains the basis for the most modern forms of navigation in the world today.

The Truth About Impotence 2510
May 12, 1998
Producer(s): Bob Burns, NOVA
NOVA reports on new hope for victims of erectile dysfunction, also known as impotence. Among the promising therapies covered in the program are ones developed by Dr. Irwin Goldstein of Boston University School of Medicine and Dr. Harin Padma-Nathan, director of the Male Clinic in Santa Monica, CA. Actual cases are profiled, featuring men talking candidly about their problem—and going through treatment. Erectile dysfunction affects an estimated 52% of men between the ages of 40 and 70.

Crocodiles with David Attenborough 2509
April 28, 1998
Producer(s): Stephen Sweigart, NOVA; Karen Bass, BBC
An unprecedented look at a dangerous predator, this natural history program is hosted by Sir David Attenborough. Surviving virtually unchanged since the days of the dinosaur and found throughout the world, these remarkable creatures have the tools for survival. Long known as vicious hunters, new photographic techniques reveals them cooperating with each other and protecting their families. From tiny babies hatching from the shell they grow into great beasts capable of standing up to a lion and bringing down a zebra.

Warnings From the Ice 2508
April 21, 1998
Producer(s): Julia Cort & Rob Gardner, NOVA
Could the world be facing the next deluge—a catastrophic rise in sea levels—as a result of the rapid break-up of the huge Antarctic ice sheets? The ice sheets hold 70% of the world's fresh water in a deep freeze cold enough to shatter steel, but now scientists are racing to understand whether the recent calving of a Connecticut sized iceberg signals the beginning of a giant meltdown.

Search for the Lost Cave People 2507
March 31, 1998
Producer(s): Susan Kopman-Lewis, NOVA; Antoine de Maximy, Gedéon
NOVA follows an international team of archeologists and spelunkers into the Rio la Venta Gorge deep in the Chiapas jungle of Central America. In a rugged canyon they find caves filled with startling remains of a people called the Zoque who lived hundreds of years before the Maya. The extreme inaccessibility and relative dryness of the caves has preserved rare artifacts including bones, clothes, rope, and jewelry. Moving downstream from the caves the team finds a legendary city hidden in a tangle of jungle vines. Evidence of the Zoque's sophisticated writing system and their practice of ritualistic cannibalism and child sacrifice is shedding new light on a little known civilization.

The Brain Eater 2505
February 10, 1998
Producer(s): Joe McMaster, NOVA; Bettina Lerner, BBC
In this scientific mystery, NOVA ventures to the front lines of medical research where scientists are scrambling to understand the strange new ailment popularly known as "mad cow disease." Highly infectious and incurable, this disease has claimed the lives of nearly a million cattle in Britain, and a variant is responsible for a handful of deaths in humans. Millions more people may have been exposed, and now the

race is on to determine if we are on the brink of another deadly epidemic like AIDS or Ebola. What scientists are finding is making them rethink many fundamental assumptions about epidemiology and may hold startling implications for public health in the future.

Supersonic Spies 2503
January 27, 1998
Producer(s): Stephen Sweigart, NOVA; Katherine Bailey, Channel 4
The race to build the world's first supersonic passenger airliner led to a massive espionage effort during the Cold War between the Soviet Union and the West. The Soviets started years behind the Concorde team, but espionage enabled Konkordski to beat French Concorde into the air by three months. Now, NOVA reveals the cause behind the fatal Konkordski disaster at the 1973 Paris Air Show, which put the Soviet's work on the plane in a deep freeze. In a twist of fate, Konkordski is being resurrected in a NASA initiative to build the second generation of supersonic jets.

Mysterious Mummies of China 2502
January 20, 1998
Producer(s): Howard Reid, NOVA & Channel 4
Perfectly preserved 3000-year-old mummies have been unearthed in a remote Chinese desert. They have long, blonde hair and blue eyes, and don't appear to be the ancestors of the modern-day Chinese people. Who are these people and how did they end up in China's Takla Makan desert? NOVA takes a glimpse through a crack in the door of history, to a past that has rarely been seen outside of China.

Night Creatures of the Kalahari 2501
January 6, 1998
Producer(s): Stephen Sweigart, NOVA; Ken Oake, Partridge Films Ltd.
Beneath the grassland plains of the Kalahari lies a hidden world of rare and exotic animals. By day, the Kalahari belongs to familiar pred-

ators and grazing animals. At night, the earth seems to release scores of seldom seen nocturnal creatures—bush babies, brown hyenas, aardvarks and fungal termites—in search of food.

Danger in the Jet Stream 2419
December 2, 1997
Producer(s): Garfield Kennedy & Peter Williams, NOVA
NOVA covers the latest efforts to be first to circumnavigate the planet non-stop in a balloon. NOVA's cameras are on board for all three attempts, including that of the long-shot underdog, American Steve Fossett, who rode high-speed winds solo from Missouri to a remote corner of India against incredible odds.

Avalanche! 2418
November 25, 1997
Producer(s): Beth Hoppe & Jack McDonald, NOVA
Viewers see what it's like to be overwhelmed by a sudden onslaught of "white death"—an avalanche. Avalanches are an escalating peril as skiers and snowmobilers push the limits into the back country. NOVA witnesses scientists getting buried alive in their attempts to understand these forces of nature.

Treasures of the Sunken City 2417
November 18, 1997
Producer(s): Julia Cort, NOVA; Max Soussana, Gedeon
Divers search for one of the seven wonders of the ancient world: the Lighthouse of Alexandria, which was destroyed in an earthquake in 1375 and some believe lies in rubble on the sea floor. Close inspection of submerged ruins reveals some monumental archaeological surprises.

Super Bridge 2416
November 12, 1997
Producer(s): Neil Goodwin, NOVA
Viewers are sidewalk supervisors for one of the most unusual construction projects in the U.S.—the building of the stunningly beau-

tiful and eminently practical Clark Bridge over the Mississippi River. Contractors faced every obstacle in the book—and then some—to build this complex structure.

Wild Wolves 2415
November 11, 1997
Producer(s): Mike Salisbury, BBC
Sir David Attenborough hosts a never-before-seen look at one of the most misunderstood creatures in nature. Unique photography, including infra-red photography, exposes the secret life of the wolf pack.

The Proof 2414
October 28, 1997
Producer(s): John Lynch, BBC
In a tale of secrecy, obsession, dashed hopes, and brilliant insights, Princeton math sleuth Andrew Wiles goes undercover for eight years to solve history's most famous math problem: Fermat's Last Theorem. His success was front-page news around the world. But then disaster struck.

Bomb Squad 2413
October 21, 1997
Producer(s): David Dugan, NOVA & Channel 4
IRA terrorists and British bomb disposal experts tell behind-the-scenes stories of a deadly cat-and-mouse game that pits ingenious IRA explosives officers against the most creative bomb squad in the world.

Faster Than Sound 2412
October 14, 1997
Producer(s): Melanie Wallace, NOVA; Tony Stark, Channel 4
On the 50th anniversary of the first supersonic flight, Chuck Yeager relives his gutsy assault on the sound barrier and tells how it was done. Other top test pilots of the day—those who survived—describe the dangers, mysteries, and thrill of trying to fly faster than sound at the dawn of the jet age.

Coma 2411

October 7, 1997

Producer(s): Linda Garmon, NOVA

A famous brain surgeon struggles to save the life of a comatose child using a controversial new method of treating severe head injuries. In charge is Dr. Jam Ghajar, who gained notoriety in 1996 by successfully treating a woman who was savagely beaten in Manhattan's Central Park and expected to die. Dr. Ghajar believes the measures that helped save her life should be available to all.

Kingdom of the Seahorse 2410

April 15, 1997

Producer(s): Susan Kopman Lewis, NOVA; Andrew Thompson, BBC

NOVA explores the secret realm and kinky sex life of one of nature's most remarkable creatures—a magical-looking fish that puts the responsibility of pregnancy and birth on the male.

Cut to the Heart 2409

April 8, 1997

Producer(s): Julia Cort, NOVA; John Hayes Fisher & Emma Walker, BBC

Would you let a part-time farmer, part-time surgeon cut out a living piece of your heart? NOVA covers a controversial surgical technique that could be a breakthrough in treating heart failure.

Curse of T. Rex 2408

February 25, 1997

Producer(s): Mark Davis, NOVA

The king of the dinosaurs is back, as public and private interests battle for the finest T.Rex specimen ever found. NOVA covers the fight for America's disappearing fossil legacy.

Hunt for Alien Worlds 2407
February 18, 1997
Producer(s): Stephen Sweigart, NOVA; Danielle Peck, BBC
All eyes are on the heavens in search for planets around other stars, probably the best hope for showing that we may not be alone in the universe. NOVA covers an effort that is turning up more and more new worlds.

Secrets of Lost Empires: Obelisk 2405
February 12, 1997
Producer(s): Michael Barnes, NOVA; Cynthia Page, BBC
As the special miniseries "Secrets of Lost Empires" continues, NOVA tests theories on how Egyptian obelisks were raised.

Secrets of Lost Empires: Colosseum 2406
February 12, 1997
Producer(s): Julia Cort, NOVA; Cynthia Page, BBC
As the special miniseries "Secrets of Lost Empires" continues, NOVA tests theories on how a gigantic awning was unfurled over the Roman Colosseum.

Secrets of Lost Empires: Inca 2404
February 11, 1997
Producer(s): Michael Barnes, NOVA; Cynthia Page, BBC
In part two of its miniseries "Secrets of Lost Empires", NOVA tackles the mysteries of how the Incas built their dramatic grass suspension bridges.

Secrets of Lost Empires: Stonehenge 2403
February 11, 1997
Producer(s): Julia Cort, NOVA; Cynthia Page, BBC
In part one of its miniseries "Secrets of Lost Empires", NOVA tackles the mysteries of how ancient Britons erected Stonehenge.

Titanic's Lost Sister 2402
January 28, 1997
Producer(s): Chuck Passerman, Kirk Wolfinger & Lisa Wolfinger, NOVA

With oceanographer Robert Ballard, NOVA dives to the watery grave of Titanic's nearly identical twin, Britannic. The luxury liner—serving as a hospital ship—was supposed to be even more unsinkable than Titanic, but she went down during World War I in near-record time under mysterious circumstances. NOVA looks for clues to explain the disaster.

Kaboom! 2401
January 14, 1997
Producer(s): David Dugan, NOVA & Channel 4

Explosions fill the screen as NOVA probes the ultimate in chemical reactions. The program explores the tarnished history of explosives from the invention of gunpowder in China to Alfred Nobel's bright idea of dynamite to the most elegant refinements in the art of blowing things up.

Cracking the Ice Age 2320
December 31, 1996
Producer(s): Joe McMaster, NOVA; David Malone & David Paterson, BBC

NOVA investigates an intriguing idea on the origin of the Ice Age: namely, the Himalayas did it. According to the theory, the crash of continents that produced Mount Everest also produced a complicated chain of effects that has resulted in a drastically altered world climate.

Odyssey of Life: Part III, The Photographer's Secrets 2319
November 26, 1996
Producer(s): Thomas Friedman, NOVA; Michael Agaton & Lars Rengfeld, SVT

Part three of the NOVA miniseries Odyssey of Life shows just what it takes to be a microphotographer like Lennart Nilsson: genius and perfection. Allowing himself to be photographed on the job for the first time, Nilsson demonstrates how he films such spectacles as a human egg cell during conception, a journey through the aorta, a kiss from a molar's point of view and the quavering vocal chords of famed opera diva Birgit Nilsson.

Odyssey of Life: Part II, The Unknown World 2318
November 25, 1996
Producer(s): Thomas Friedman, NOVA; Michael Agaton & Lars Rengfeld, SVT

Part two of the NOVA miniseries Odyssey of Life spotlights the tiny monsters that live in, on and around us. Noted microphotographer Lennart Nilsson shows how a tame-looking lawn is, in fact, a horror movie of battling bugs, a closet of clothes is a cafeteria of lunching larvae and glamorous eyelashes are a gallery of munching mites—if you look closely enough.

Odyssey of Life: Part I, The Ultimate Journey 2317
November 24, 1996
Producer(s): Thomas Friedman, NOVA; Michael Agaton & Lars Rengfeld, SVT

Showcasing the work of renowned microphotographer Lennart Nilsson, the first of a three-part series shows how four billion years of evolutionary history are reflected in life before birth—in species as different as humans, chickens and fish. Along the way, Nilsson captures the beautifully simple organisms that gave rise to complex life forms and shows how evolution often works in surprising ways.

Shark Attack! 2316
November 19, 1996
Producer(s): Susan Kopman Lewis, NOVA; Doug Bertran & Petra Regent, Survival Anglia

NOVA explores the awesome power of the most dangerous hunter of the deep. Traveling to shark-infested waters off California and the

Hawaiian Islands, the program presents breathtaking footage of sharks attacking their natural prey—and looks at the chances of humans becoming their catch of the day.

Top Gun Over Moscow 2315
November 12, 1996
Producer(s): Lance K. Shultz & Lynne Squilla, NOVA
NOVA lights up the afterburners to celebrate the meeting of former adversaries from the Cold War. Russian pilots host American jet jockeys to show off the hottest planes from the former Soviet Union, which prove surprisingly rugged compared with their pampered US counterparts.

Secrets of Making Money 2314
October 22, 1996
Producer(s): Susan Kopman Lewis, NOVA
NOVA inspects the newly redesigned $100 bill which includes a panoply of hard-to-reproduce features aimed at thwarting the growing ranks of counterfeiters, including the new breed who multiply their wealth at the local copy shop.

Three Men and a Balloon 2313
October 15, 1996
Producer(s): Garfield Kennedy & Peter Williams, NOVA
NOVA takes viewers on the ride of their lives, a trouble-plagued attempt by three wealthy adventurers to be the first to ride a balloon non-stop around the world. Will this marriage of three overpowering egos beat the competition for the last great prize in aviation?

Lost City of Arabia 2312
October 8, 1996
Producer(s): Nicholas Clapp, Tom Friedman & George R. Hedges, NOVA
NOVA transports viewers to the shifting sands of Arabia on a

quest for the fabled lost city of Ubar. Featured in the Arabian Nights , this ancient caravan stop has eluded explorers for many centuries, until data from the Space Shuttle helped pin down it's location.

Einstein Revealed 2311
October 1, 1996
Producer(s): Peter Jones & Tom Levenson, NOVA
Einstein is explored and explained in a two-hour special probing the genius and foibles of the great scientist as revealed in his personal papers. Actor Andrew Sachs depicts Einstein being interviewed in his Berlin study on the eve of the Nazi takeover in 1932. Plus, the latest computer animation gives the most lucid explanation ever of the theory of relativity.

Alive on Everest Special
May 17, 1996
Producer(s): Liesl Clark, NOVA
"Alive on Everest" combines breaking news from the mountain with a compelling documentary that gives viewers a first hand look at what it takes to climb the world's highest-and deadliest-peak. The film includes an interview with David Breashears direct from Mt. Everest followed by the personal story of New Englander Ric Wilcox's dramatic 1991 attempt to reach the mountain's summit.

Warriors of the Amazon 2309
April 29, 1996
Producer(s): Melanie Wallace, NOVA; Chris Curling, Channel 4
NOVA travels to the Amazonian jungle to live among the Yanomami, one of the few remaining hunter/gatherer groups in the world, recording their healing ceremonies, death practices and other customs, including a ritual feast with their enemies.

Dr. Spock the Baby Doc 2308
April 2, 1996
Producer(s): Robert Richter, NOVA
NOVA profiles Dr. Benjamin Spock, whose best-selling baby and child care guide revolutionized the way Americans raise their children. At ninety-something, Dr. Spock continues to mix a lively interest in babies with his long-standing activism for world peace, on the theory that war is potentially more dangerous to children than accidents or illness.

Flood! 2307
March 26, 1996
Producer(s): Larkin McPhee, NOVA
In 1993 the Mississippi River swept away bridges, levees, farms and entire towns in the largest flood ever recorded in America's heartland. NOVA covers the human drama of the flood-fight to stop a river overflowing from weeks of nearly nonstop rain.

Bombing of America (The) 2310
March 26, 1996
Producer(s): Nancy Linde, NOVA
NOVA examines what science and technology are doing to help law enforcement officials battle the bombing epidemic that is on the rise in America. To show what's involved, NOVA profiles several notorious bombing incidents, including the celebrated case of the Unabomber and the eerily similar career of the "Mad Bomber" of New York.

War Machines of Tomorrow 2305
February 20, 1996
Producer(s): Larry Klein, NOVA
NOVA travels to the testing ranges and training grounds for a leaner, meaner and more effective United States military force that can fight and win on almost any battlefield in the world. One innovation in the works: super-accurate "brilliant" weapons, designed as successors to the smart munitions used in the Gulf War.

Plague Fighters 2304
February 6, 1996
Producer(s): Ric Esther Bienstock, Elliott Halpern & Simcha Jacobovici, NOVA
In the spring of 1995 a deadly outbreak of the Ebola virus swept through Kikwit, Zaire, killing 77 percent of those who fell ill. No one stayed in the infectious "hot zone" longer than NOVA's production team which filmed the inside story of the battle to contain one of the most feared diseases on the planet.

B-29 Frozen in Time 2303
January 30, 1996
Producer(s): Mike Rossiter, NOVA
NOVA accompanies famed test pilot Darryl Greenamyer and his intrepid crew on a perilous mission to repair and re-fly a B-29 bomber stranded on the Greenland icecap since 1947. In the face of incredible hardships, the team struggles to bring the old warbird back to life.

The Day the Earth Shook 2302
January 16, 1996
Producer(s): Simon Campbell-Jones & Suzanne Campbell-Jones, NOVA
On the same day in January one year apart, earthquakes of almost identical power shook Northridge, California (1994) and Kobe, Japan (1995). NOVA probes why almost 100 times more people died in Japan than in the United States and what scientists learned from the twin calamities.

Terror in the Mine Fields 2301
January 9, 1996
Producer(s): David Feingold, NOVA
In one-third of the world's countries, a misstep can mean a lost leg—to a landmine. Planted during an ongoing conflict or a war long since over, these invisible weapons lurk, ready to explode at any time. NOVA's unprecedented access to the elusive Khmer Rouge in Cam-

bodia reveals the ease of laying mines and the difficulty and danger of clearing them.

Can Buildings Make You Sick? 2217
December 26, 1995
Producer(s): Melanie Wallace, NOVA; Mike Tomlinson, BBC
NOVA unravels baffling cases of bad air in buildings all over the world. Even hospitals are on the "sick-building" list—along with offices, schools, homes and just about any enclosed space.

Race to Catch a Buckyball 2216
December 19, 1995
Producer(s): Susan Kopman, NOVA; John Lynch, BBC
Considered by many scientists the discovery of the decade, "buckyballs" are soccer-ball-shaped carbon molecules with many remarkable properties. NOVA tells how astronomers looking for the secrets of stardust came across these miraculous molecules.

Treasures of the Great Barrier Reef 2215
November 28, 1995
Producer(s): Neil O'Hare & Melanie Wallace, NOVA
Recording sights that will astonish even experienced divers, NOVA documents an extraordinary day in the life of the largest coral reef in the world, capturing the rarely seen annual spawning of coral and other unusual creatures of the reef.

Hunt for the Serial Arsonist 2214
November 14, 1995
Producer(s): Carl Charlson, NOVA
Who's behind the rash of suspicious fires in Los Angeles? NOVA hits the streets with arson sleuths who nail a surprising suspect. Like a fire raging out of control, this case takes many odd twists and turns.

Lightning! 2213
November 7, 1995
Producer(s): Linda Garmon, NOVA
NOVA profiles nature's spectacular light show in electrifying detail, getting up close as scientists provoke lightning strikes to learn more about them. The program also looks at lightning's survivors, who were shocked to find themselves victim to nature's fury.

The Doomsday Asteroid 2212
October 31, 1995
Producer(s): Lauren S. Aguirre, NOVA; Tim Haines, BBC
Is there an asteroid or comet out there with our name on it? NOVA scans the skies and the geological record on Earth, for evidence that giant rocks from outer space have struck before and will eventually plow into our planet again.

Hawaii Born of Fire 2211
October 24, 1995
Producer(s): Neil Goodwin, NOVA
NOVA explores the fiery moonscapes and lush rainforests of the world's most isolated archipelago: the Hawaiian islands. From blistering beginnings as molten rock, the islands have developed into a verdant paradise of unique lifeforms

Venus Unveiled 2210
October 17, 1995
Producer(s): Joseph McMaster, NOVA; Bettina Lerner, BBC
Venus reveals its true face, recorded in detail for the first time by the radar spacecraft Magellan. Our next-door planetary neighbor turns out to be one of the most bizarre places in the solar system.

Anastasia Dead or Alive 2209
October 10, 1995
Producer(s): Michael Barnes, NOVA
Did a daughter of the last czar of Russia escape her family's bloody

fate and end up living in Charlottesville, Virginia? From bones recently uncovered to testimony from royal family insiders, NOVA presents new evidence about what really happened to Anastasia, youngest daughter of czar Nicholas.

Fast Cars 2208
May 23, 1995
Producer(s): Sam Low, NOVA
"All of racing is trying to get a little something more than someone else," says racing veteran and Indy car team owner Carl Hogan. NOVA follows the trials and tribulations of Hogan and his co-owner and driver, former Indianapolis 500 winner Bobby Rahal, as they try to engineer a car that can edge out the competition.

Making of a Doctor 2207
May 3, 1995
Producer(s): Michael Barnes, NOVA
NOVA completes its third and final installment of an unprecedented ten-year project to chronicle the complete medical education of a group of Harvard Medical students—from day one of medical school, through internships and residencies.The participants in this decade-long documentary are Elliott Bennett-Guerrerro, Jay Bonnar, Cheryl Dorsey, Luando Grazette, David Friedman, Jane Liebschutz and Tom Tarter.

Universe Within, The 2206
March 7, 1995
Producer(s): Beth Hoppe, NOVA; Bo. G. Erikson, Katuhiko Hayashi & Carl O. Lofman, NHK & SVT
As superstar athletes push their bodies to the limit, NOVA takes an amazing journey inside to see what makes it all possible—watching as the body turns junk food into raw energy, experiencing one climber's struggle to survive the onslaught of infection, and getting up close and personal as sperm meets egg.

Kidnapped by UFO's? 2306
February 27, 1995
Producer(s): Denise DiIanni, NOVA
Thousands of Americans have come forward with tales of being kidnapped by space aliens, and millions of Americans believe them. NOVA searches for the truth behind real-life stories, worthy of "The X Files", describing late-night visits by small, gray creatures bent on creating a hybrid human/alien race.

Mystery of the Senses—part 4 Touch MYSS 104
February 22, 1995
Producer(s): Peter Jones, NOVA
On the fourth episode of Mystery of the Senses hosted by Diane Ackerman, NOVA explores our most sensual sense, and learns that touching is a potent tonic, not only for caressing others, but for premature infants who thrive when massaged. Touch also serves as a tool for artists and as "eyes" for the blind.

Mystery of the Senses—part 5 Vision MYSS 105
February 22, 1995
Producer(s): Peter Jones, NOVA

Mystery of the Senses—part 3 Taste MYSS 103
February 21, 1995
Producer(s): Peter Jones & Thomas Levenson, NOVA
In the third episode of Mystery of the Senses hosted by Diane Ackerman, NOVA shows how our basic need for nourishment has led to the miracle of great cooking and eating. A two-star meal in France, a traditional Mexican feast designed to nourish the dead, and the world's best street food are some of the enlightening eats in this gustatory odyssey.

Mystery of the Senses—part 2 Smell MYSS 102
February 20, 1995
Producer(s): Michael Gunton, Peter Jones & Larry Klein, NOVA

Exotic smells almost waft from the TV as viewers visit the world's largest perfumery in the second episode of Mystery of the Senses hosted by Diane Ackerman. NOVA takes viewers beyond the fragrance factory to sample a huge spectrum of smells-from frankincense and truffles to sweaty T-shirts and worse. Interestingly, all share elements that are irresistibly attractive.

Mystery of the Senses—part 1 Hearing MYSS 101
February 19, 1995
Producer(s): Peter Jones, Larry Klein, NOVA & WETA

In the premiere episode of Mystery of the Senses hosted by Diane Ackerman, NOVA takes viewers on a kaleidoscope journey to "hear" the world around us; to the quietest place on earth, to hear how movie sound effects are made, to experience the music-rich culture of the Maori of New Zealand and to accompany a deaf woman as she lives through an operation that might restore her hearing.

Siamese Twins 2205
February 14, 1995
Producer(s): Jon Palfreman, NOVA

Dao and Duan are literally Siamese twins, for the land where they were born is Thailand, formerly called Siam. The orphan girls were sent to America in April 1993 to be evaluated at The Children's Hospital of Philadelphia, where a surgical team led by Dr. James O'Neill has performed more than a dozen successful separations of conjoined twins. NOVA follows the saga that turn the inseparable duo into distinct individuals.

Nazi Designers of Death 2204
February 7, 1995
Producer(s): Isabelle Rosin & Mike Rossiter, NOVA
The discovery of top-secret Nazi files reopens a painful chapter in history, revealing the careful planning behind the Nazi death camps. Using recently discovered documents, NOVA uncovers the complicity of the professional engineers and architects who turned the practical technology of their craft to wholesale slaughter.

Little Creatures Who Run the World 2203
January 31, 1995
Producer(s): Peter Jones & Nick Upton, NOVA
Humans may think they run the world, but there's another superpower who is really on top. They outnumber us a million to one—and little can stand in their way. Their engineers breach wide gaps in a single bound. Their workers lift weight twice their size. Their soldiers are studied by U.S. defense analysts.Who is this superpower, and what makes them so successful? NOVA gets up close and personal with ants, the little creatures who run the world.

Vikings in America 2202
January 24, 1995
Producer(s): Thomas Friedman & T.W. Timreck, NOVA
Five hundred years before Columbus, the Vikings (also known as Norseman) reached North America. Who were the people they met here? What happened when two worlds collided? Archeologists are now revealing an extraordinary story of tragedy and triumph. NOVA investigates the myth and reality of the first known Europeans to reach North America.

Mammoths of the Ice Age 2201
January 9, 1995
Producer(s): Sheila Hairston, NOVA ; Kate O'Sullivan, BBC
Ten thousand years ago, a world frozen in ice began to thaw, marking the beginning of the end for the great woolly mammoth. But what ef-

fect did humans have on these huge creatures? Frozen bodies and houses made of tusks are just some of the amazing new finds. Now, scientists are piecing together a picture of the life our ancestors shared with the woolly mammoth.

In Search of the First Language 2120
December 27, 1994
Producer(s): Melanie Wallace, NOVA; Christopher Hale, BBC
NOVA explores the common threads that link the more than 5,000 languages of the globe, including a controversial theory that claims to link all languages to a common ancestral "Mother Tongue", recalling the biblical story of the Tower of Babel.

Journey to the Sacred Sea 2119
December 20, 1994
Producer(s): Thomas Friedman, NOVA; Frances Berrigan, Channel 4
NOVA travels to Lake Baikal, the world's oldest and deepest lake, containing one-fifth of all the fresh water on Earth. Investigating Baikal from above, below and all around, NOVA charts its dramatically changing environment over the course of four seasons.

Rescue Mission in Space 2118
December 13, 1994
Producer(s): Lauren S. Aguirre, NOVA
Hobbled by defective eyesight because of its original, bungled prescription, the Hubble Space Telescope was recently repaired in a dramatic Space Shuttle mission. NOVA follows the exploits of astronauts who saved the day, and the stunning work that Hubble has performed in the months since its repair.

Buried in Ash 2117
November 29, 1994
Producer(s): Lisa Mirowitz, NOVA; Gary Hochman, Nebraska ETV
Ten million years ago, an enormous volcanic eruption buried much of what is now Nebraska in up to ten feet of ash, preserving countless skeletons of prehistoric big game animals. NOVA joins paleontologist and Nebraska native, Mike Voorhies, the discoverer of this treasure trove, to learn what life was like when the West was filled with big game animals.

Killer Quake! 2116
November 15, 1994
Producer(s): Robert Dean, NOVA
NOVA probes the 1994 Los Angeles earthquake. Even as the city struggles to repair itself from the tragedy, seismic pressure continues to build. Scientists fear that newly discovered faults could, at any moment, trigger California's most devastating natural disaster.

Tribe That Time Forgot (The) 2115
November 8, 1994
Producer(s): John Miles, NOVA
NOVA travels deep into the Amazon wilderness in search of a mysterious tribe that dismembered and partially ate three prospectors in 1976. Locating the group, NOVA lives with them for three months, gaining insight into the customs and beliefs of a people whose lifestyle have changed little over centuries.

What's New About Menopause 2114
November 1, 1994
Producer(s): Denise DiIanni, NOVA
NOVA tackles the long-taboo subject of menopause, profiling new research and examining the medical and ethical controversies that arise when science enables women to postpone menopause or even to bear children long after "the change." Stockard Channing narrates.

Haunted Cry of a Long Gone Bird 2113
October 25, 1994
Producer(s): David Conover & Christopher Knight, NOVA
NOVA explores the legacy of the great Auk, a magnificent flightless bird that was hunted to extinction over a century ago. In a journey retracing its migratory route, host Richard Wheeler kayaks from Newfoundland to Cape Cod and discovers that other marine species are facing the Auk's luckless fate.

Secret of the Wild Child 2112
October 18, 1994
Producer(s): Linda Garmon, NOVA
NOVA profiles "Genie," a girl whose parents kept her imprisoned in near total isolation from infancy. Includes footage of Genie during her rehabilitation and probes how and when we learn the skills that make us "human."

Great Wildlife Heist (The) 2111
October 11, 1994
Producer(s): Al Austin & Susan Lambert, NOVA, OPB & Film Australia
NOVA goes undercover with a U.S. government sting that breaks an international parrot smuggling ring, landing some surprising suspects.

Aircraft Carrier 2110
April 19, 1994
Producer(s): Mike Rossiter, NOVA
NOVA experiences the relentless, round-the-clock life aboard the U.S. Navy aircraft carrier, Independence where every day is a constant drill of launching and landing aircraft atop a floating city of 5,000 people. The action includes Top Gun mock combat exercises and live-ammunition patrols over Iraq.

Can China Kick the Habit? 2109
April 12, 1994
Producer(s): Sue DeMarco & Judy Katz, NOVA; Gill Barnes, Channel 4
NOVA visits the most cigarette-addicted nation in the world—China. Western advertising and trading practices have exacerbated the fatal romance with smoking in the world's most populous country, where lung cancer cases are beginning to strain the nation's health care system.

In Search of Human Origins: Part Three 2108
March 2, 1994
Producer(s): Lauren Seeley Aguirre & Peter Jones, NOVA
At what point did our distant ancestors become anatomically like us? And, more importantly, when did they begin to act like us? Paleoanthropologist Donald Johanson looks at what it is that makes us human.

In Search of Human Origins: Part Two 2107
March 1, 1994
Producer(s): Lenora Johanson & Peter Jones, NOVA
Paleoanthropologist Donald Johanson looks at how our human ancestors of two million years ago made their living. Contrary to popular myth, scavenging was a more lucrative living than hunting—and may have contributed to the development of human intelligence.

In Search of Human Origins: Part One 2106
February 28, 1994
Producer(s): Michael Gunton & Peter Jones, NOVA
In the first of a three-part series, noted paleoanthropologist Donald Johanson probes the earliest ancestors of the human species—reaching back more then 3 million years to a strange ape who walked upright. Johanson takes viewers to the site in Ethiopia where he discovered the fossil remains of this missing link nicknamed "Lucy."

Can Chimps Talk? 2105
February 15, 1994
Producer(s): Masaru Ikeo, NHK
NOVA covers exciting and controversial research with chimpanzees who have been trained to express themselves with human symbols. Are they speaking their minds, or are they just aping their trainers?

Journey to Kilimanjaro 2104
February 8, 1994
Patrick Morris, BBC
Rising straight out of the parched plains of East Africa are colossal ice-capped mountains. Towering up to four miles high, these volcanic peaks are a hostile world of freezing nights, scorching days and smoking summits. NOVA explores this unique archipelago of life, traveling from an equatorial blizzard on Mount Kenya, through secret cloud forests in the Aberdares range, to the majestic crown of mighty Kilimanjaro.

Daredevils of the Sky 2103
February 1, 1994
Producer(s): Charles Passerman & Kirk Wolfinger, NOVA
NOVA follows members of the U.S. Aerobatic Team as they prepare for and compete in the 1992 World Aerobatic Championship. The program also tells the story of aircraft designer Jon Staudacher, who is working on the next generation of high-performance aerobatic planes.

Dinosaurs of the Gobi 2102
January 25, 1994
Producer(s): Susan Kopman, NOVA; Anthony Edwin Nahas, BBC
NOVA accompanies an American Museum of Natural History expedition to the Gobi Desert. The trip relives the exploits of the Museum's dashing explorer of the 1920s, Roy Chapman Andrews—said to be the real-life model for Indiana Jones.

Codebreakers 2101
January 18, 1994
Producer(s): William Woolard, NOVA
NOVA delves into the history of secret communications and the people who decipher them, probing the most celebrated of all cryptographic coups: the breaking of the World War II codes used by Japan and Germany, and how codebreaking helped shorten the war.

Stranger in the Mirror 2020
December 28, 1993
Producer(s): Alain Jehlen, NOVA; Hilary Lawson, BBC
NOVA explores the nature of human perception through the puzzling condition called visual agnosia, the inability to recognize faces and familiar objects, made famous in Oliver Sacks' book, The Man Who Mistook His Wife for a Hat.

Best Mind Since Einstein, The 2019
December 21, 1993
Producer(s): Melanie Wallace, NOVA; Christopher Sykes, BBC
A profile of the late Richard Feynman—atomic bomb pioneer, Nobel prize-winning physicist, acclaimed teacher and all-around eccentric, who helped solve the mystery of the space shuttle Challenger explosion.

Great Moments From NOVA 2018
December 7, 1993
Producer(s): Beth Hoppe, NOVA
Bill Cosby guides viewers through the most exciting footage from two decades of NOVA in a 20th anniversary salute. Real-life action, adventure, mystery, drama and non-stop discovery fill this 90-minute special.

Mysterious Crash of Flight 201 2017
November 30, 1993
Producer(s): Marian Marzynski, NOVA
The USA's crack team of investigators from the National Transportation Safety Board are called in to figure out the mysterious cause of a jetliner crash in Panama. Nothing about the accident makes sense, until a key clue emerges.

Roller Coaster! 2016
November 16, 1993
Producer(s): Thomas Freidman & Larkin McPhee, NOVA; Frankie Glass, Channel 4
NOVA takes viewers on the ride of their lives as it explores the science of roller coasters, where physics and psychology meet. New rides of the future may take place entirely in the mind—with virtual reality.

Real Jurassic Park (The) 2015
November 9, 1993
Producer(s): David Dugan, NOVA
With help from director Steven Spielberg, author Michael Crichton and a host of scientific experts, NOVA investigates what it would take to recreate the dinosaur theme park in Jurassic Park.

Shadow of the Condor 2014
November 2, 1993
Producer(s): Tom Friedman, NOVA; Olivier Bremond, Marathon TV, France
NOVA soars with the condor, an extraordinary bird that lives a tenuous existence in the California mountains and the Andes of South America. Footage includes never-before-photographed nesting sites in the cliffs of Patagonia.

Dying to Breathe 2013
October 26, 1993
Producer(s): Elias Petras & John Spotten, National Film Board of Canada
NOVA covers the tense vigil of three people with terminal lung disease as they await the most complex of all organ transplants—a new lung. Months of waiting end in a few frenzied hours of intricate surgery.

Secrets of the Psychics 2012
October 19, 1993
Producer(s): Carl Charlson, NOVA
Magician James "The Amazing" Randi tests the claims of mind readers, fortune tellers, faith healers and others with purported paranormal powers.

Wanted: Butch and Sundance 2011
October 12, 1993
Producer(s): David Dugan, NOVA
Forensic sleuth, Clyde Snow, and a pose of experts travel to Bolivia in search of the remains of Butch Cassidy and the Sundance Kid. They find Hollywood and legend got a few things wrong.

NOVA Quiz (The) 2010
October 5, 1993
Producer(s): Nancy Linde, NOVA
NOVA fans from around the country match wits in a fast-paced contest of general science knowledge. Famous guests pose questions for the viewers at home. Marc Summers hosts.

The Lost Tribe 2009
March 30, 1993
Producer(s): Bill Lattanzi, NOVA; Bettina Lerner, BBC
NOVA covers both sides of the stormy controversy over the Tasaday tribe. When these isolated cave dwellers were discovered in the

Philippines in 1971, they were hailed as a Stone Age relic. Now, many anthropologists denounce them as fakes.

Murder, Rape and DNA 2008
March 2, 1993
Producer(s): Stephen Jimenez, NOVA; Derek Braithwaite, Central Independent Television
Inside every human cell is a set of instructions, coded on strands of DNA, that defines us uniquely as individuals. Wherever we leave these microscopic cells behind—blood smear, strand of hair—we leave an indisputable calling card of our presence. NOVA investigates DNA typing, a powerful new crime-solving tool from the frontiers of biotechnology and how this technique of molecular study is not only securing convictions, but exonerating innocent people as well.

Diving for Pirate Gold 2007
February 23, 1993
Producer(s): Larry Engel, NOVA
Today a new breed of pirates is plundering the watery remains of the old swashbuckling culture. Using sonar sounders, magnometers and metal detectors, these latter-day treasure hunters may be irretrievably destroying evidence from the era of clashing cutlasses and pieces of eight. NOVA travels to Jamaica, the Bahamas, the Florida Keys and Cape Cod to investigate the pirates of yesteryear and their modern high-tech successors.

Can Science Build a Champion Athlete? 2006
February 16, 1993
Producer(s): Richard Sattin, NOVA
Increasingly, athletes of every stripe are training smarter and performing better thanks to high technology. NOVA covers this record-setting trend in champions past, present and potential. With every possible way of improving athletic performance, records seem destined to keep falling. Now coaches can analyze numbers for any

event, and they know how to use this information to achieve peak performance in their athletes.

In the Path of a Killer Volcano 2005
February 9, 1993
Producer(s): Noel Buckner & Rob Whittlesey, NOVA
Covering the largest volcanic eruption in 80 years, NOVA follows scientists as they monitor the increasingly ominous signs from Mt. Pinatubo. While keeping local officials alerted, it is 'touch and go' for the scientists as they try to decide whether to call for an evacuation. They decide just in time before the Philippine volcano blows it's top.

Nazis and the Russian Bomb 2004
February 2, 1993
Producer(s): Larkin McPhee, NOVA; Mike Rossiter & Toni Strasburg, BBC
In May 1945, while the United States was luring Nazi rocket scientists to America, the Soviets were picking out German atomic scientists to help them in their efforts to catch up with the U.S. nuclear program, taking them prisoner to design nuclear weapons and long-range rockets. NOVA covers the astonishingly swift success of the Soviet Union to build itself into a nuclear superpower.

The Deadly Deception 2003
January 26, 1993
Producer(s): Denise DiIanni, NOVA
Forty years ago, the United States government conducted one of the most notorious medical experiments in American history: the Tuskegee Study of Untreated Syphilis in the Negro Male. Presented by George Strait, ABC News Medical Correspondent, NOVA investigates this infamous human experiment that many believe perverted principles of medical ethics and the relationship between patient and doctor and subject and researcher.

Can Bombing Win a War? 2002
January 19, 1993
Producer(s): Thomas Friedman, NOVA
NOVA charts the history of strategic bombing. The program features analysis of the bombing campaign against Iraq by US Air Force officers. Veterans of World War I, World War II and Vietnam recount their experiences with the deadly, and often ineffective, business of attacking the enemy from the air.

The Hunt for Saddam's Secret Weapons 2001
January 12, 1993
Producer(s): Sheila Hairston & Melanie Wallace, NOVA; Victoria Schultz, United Nations
NOVA reports on the international team of advisors who are fulfilling the United Nations' mandate to dismantle Iraq's weapons of mass destruction. A behind-the-scenes look at the cat-and-mouse game that UN inspectors must play with Iraqi officials.

Sex and the Single Rhino 1920
December 29, 1992
Producer(s): Melanie Wallace, NOVA; Robert Thirkell, BBC
Today's job of saving endangered species involves sophisticated computer modeling, international diplomacy and delicate laboratory techniques that have no guarantee of success. NOVA examines this high-tech effort to save future generations of rhinoceros and other rare animals.

Can You Stop People From Drinking? 1919
December 22, 1992
Producer(s): David Dugan & Eric Stange, NOVA
NOVA covers the discouraging and difficult efforts to curb alcohol abuse in the United States and Russia.

Brain Transplant 1918
December 1, 1992
Producer(s): Jon Palfreman, NOVA
NOVA follows a remarkable, little-known, medical detective story that started with an inexplicable paralysis among drug abusers. The trail led doctors from a bad batch of synthetic heroin to a research breakthrough in understanding Parkinson's Disease, and even to the prospect of curing some brain diseases with fetal implants.

Private Lives of Dolphins (The) 1917
November 17, 1992
Producer(s): Mark Davis, NOVA
NOVA follows Randy Wells (Sarasota, Florida) and Richard Connor (Shark Bay, Australia) as they try to make sense of the complex society, sex and politics of the bottle-nosed dolphin.

Iceman 1916
November 10, 1992
Producer(s): Sheila Hairston, NOVA; Katharine Everett, BBC
NOVA covers the international efforts to unlock the secrets behind the astonishing discovery of the mummified body of a man found in the Alps on the Italian/Austrian border by two German hikers. Scientists astonished the world when they reported that he was 5000 years old.

This Old Pyramid 1915
November 4, 1992
Producer(s): Michael Barnes, NOVA
NOVA reveals the ancient secrets of Pyramid building by actually constructing one, putting clever and sometimes bizarre pyramid construction theories to the test. This monument to the most astonishing engineering feat of antiquity was built under the guidance of Dr. Mark Lehner of the University of Chicago's Oriental Institute, and Roger Hopkins, a Massachusetts stonemason.

Rafting Through the Grand Canyon 1914
October 27, 1992
Producer(s): Linda Harrar, NOVA
NOVA explores the Grand Canyon—the most dramatic landform on the face of the earth—by rafting down the river that created it. Excerpts from the diary of 19th-century canyon explorer John Wesley Powell are read.

Search for the First Americans 1913
October 20, 1992
Producer(s): Evan Hadingham, NOVA; Simon Campbell-Jones, BBC
NOVA probes the mystery of America's first inhabitants—the true discoverers of the New World, travelling to several controversial sites in North and South America where evidence suggests that Native Americans may have first entered the New World thousands of years earlier than assumed by generations of archeologists.

Mind of a Serial Killer 1912
October 13, 1992
Producer(s): Larry Klein & Mark Olshaker, NOVA
NOVA goes behind the scenes and into the frontiers of modern crime detection to probe the FBI's Investigative Support Unit, the unit responsible for getting into the minds of serial killers and hunting them down.

The Genius Behind the Bomb 1911
September 29, 1992
Producer(s): Alain Jehlen & Helen Weiss, NOVA
Profile of Leo Szilard, "inventor" of the atomic bomb.

Animal Olympians 1910
August 25, 1992
Producer(s): Beth Hoppe, NOVA; Hilary Jeffkins, BBC
How marvelous are the physical achievements of our greatest ath-

letes. But even the best can't compete with the swimmers, sprinters and high-jumpers of the animal world. In this program, NOVA looks at how human athletes compare with their counterparts in the animal kingdom.

Eclipse of the Century 1909
March 24, 1992
Producer(s): Thomas Levenson, NOVA
The spectacular solar eclipse of 1991 passed over major observatories for the first time. For astronomers in the observatories of Hawaii, the 4 minute, 8 second event provided a wealth of data which promises to reveal new secrets about the sun.

Astronaut's View of Earth, An 1908
March 17, 1992
Producer(s): Joseph Blatt, NOVA
Astronauts share their best movies of Earth, including eye-popping footage from the IMAX/OMNIMAX films "The Blue Planet" and "The Dream Is Alive."

Rescuing Baby Whales 1907
March 10, 1992
Producer(s): Mark Davis, NOVA
Scientists are gradually learning to cope with mysterious whale strandings along the beaches of Cape Cod Bay off the Massachusetts coast.

Making a Dishonest Buck 1906
March 3, 1992
Producer(s): David Dugan, NOVA
Criminals still make money the old-fashioned way—by counterfeiting. NOVA looks at why US currency is so easy to fake and what the government is doing about it.

Can You Believe TV Ratings? 1905
February 18, 1992
Producer(s): Mike Tomlinson, NOVA
Measuring the audience for television shows is a classic problem in statistical analysis. NOVA finds that ratings measurement is getting more accurate but is still far from scientific.

What Smells? 1904
February 11, 1992
Producer(s): Larkin McPhee, NOVA; Vishnu Mathur, CBC
The nose knows. Just how the nose does its job is the subject of NOVA's investigation into the mysterious aromas and hidden messages picked up by our sense of smell. David Suzuki hosts.

Saddam's War on Wildlife 1903
January 28, 1992
Producer(s): Alain Jehlen & Alma Taft, NOVA
John Walsh of the WSPA (World Society for the Protection of Animals) travels to Kuwait to assist in wildlife rescue efforts in the wake of the Iraqi army's devastating impact on the environment.

Submarine! 1902
January 21, 1992
Producer(s): Charles Passerman, NOVA
NOVA prowls the deep with the super-silent ballistic missile submarine USS Michigan.

Hellfighters of Kuwait 1901
January 14, 1992
Producer(s): Mike Rossiter, NOVA
NOVA follows efforts to put out the Kuwaiti oil fires set by Iraqi soldiers during the Gulf War.

Fine Art of Faking It, The 1820
December 17, 1991
Producer(s): Denise DiIanni, NOVA
Science comes to the aid of art. Museums now employ scientists to find forgeries and give insight into the process of artistic creation.

Skyscraper! A NOVA Special 1819
December 10, 1991
Producer(s): Thomas Friedman, NOVA; Karl Sabbagh, Channel 4
NOVA follows the construction of a skyscraper in Manhattan, Worldwide Plaza, from architect's drawings to tenant occupation.

Avoiding the Surgeon's Knife 1818
December 3, 1991
Producer(s): Alain Jehlen, NOVA; Henry Corra, Maysles Films
NOVA follows the efforts of four participants in a celebrated race against death, an experimental study to unblock arteries without using drugs or surgery. Dr. Dean Ornish's study looked at the effects of drastic lifestyle changes on people with blocked coronary arteries, by asking participants to follow a strict vegetarian diet, engage in mild daily exercise and practice stress-reduction techniques.

Fastest Planes in the Sky 1817
November 12, 1991
Producer(s): John Alexanders, NOVA; Chris Haws, Channel 4
NOVA looks at the intoxicating lure to fly ever faster—from the deadly GeeBee racer of the 1930s to the X-30 "National Aero-Space Plane" (NASP) planned for the next century.

Taller Than Everest? 1816
November 5, 1991
Producer(s): Kathleen Bernhardt & David Breashears, NOVA
The tallest mountain in the world? Think again—cartographers had to re-examine their maps when satellite data revealed a peak called "K2" might be taller than Everest.

Suicide Mission to Chernobyl 1815
October 22, 1991
Producer(s): Larkin McPhee, NOVA; Edward Briffa, BBC
NOVA accompanies Soviet scientists on a deadly mission inside the sarcophagus—exploring the massive structure that entombs the Chernobyl nuclear reactor to discover the potential for another deadly explosion.

Secrets of the Dead Sea Scrolls 1814
October 15, 1991
Producer(s): Nancy Porter, NOVA
Forty years after they were discovered, the Dead Sea Scrolls have yet to be published in their entirety. NOVA looks at the laborious—some say scandalous—process of compiling and releasing this religious treasure.

So You Want to be a Doctor? Part 1 1812
October 9, 1991
Producer(s): Michael Barnes, NOVA
In a two hour special, NOVA follows seven aspiring doctors through four years of medical school. The first examination, the anatomy lab, the first death, the first baby—it's all part of becoming a doctor. Neil Patrick Harris, star of ABC's Doogie Howser, M.D., hosts.

So You Want to Be a Doctor? Part 2 1813
October 9, 1991
Producer(s): Michael Barnes, NOVA
In a two hour special, NOVA follows seven aspiring doctors

through four years of medical school. The first examination, the anatomy lab, the first death, the first baby—it's all part of becoming a doctor. Neil Patrick Harris, star of ABC's Doogie Howser, M.D., hosts.

Sex, Lies and Toupee Tape 1811
October 1, 1991
Producer(s): Joel Olicker, NOVA; Chris Haws, Channel Four
NOVA covers the causes and attempted cures of baldness. Some men take pride in their bald heads; others will go to great lengths to cover up. Alan Rachins, Douglas Brackman of NBC's LA Law, hosts.

Confusion in a Jar 1802
April 30, 1991
Producer(s): Bebe Nixon, NOVA; James Burge, BBC
On March 23, 1989, Dr. B. Stanley Pons of the University of Utah and Dr. Martin Fleischmann of the University of Southampton announced that they had created nuclear fusion in a test tube at room temperature. In the months following that announcement, scientists found it difficult to reproduce the team's results. This program examines the controversy, and the interaction of politics, business, the press and scientific research.

Chip vs. the Chess Master, The 1803
March 26, 1991
Producer(s): Irv Drasnin, NOVA
Chess is considered by many to be the ultimate game of intellectual challenge, requiring intelligence, intuition, imagination, strategy and reasoning- traits not usually associated with a computer. NOVA explores what it took to prepare Deep Thought, a computer chess program, to take on world champion Gary Kasparov in 1989.

Swimming with the Whales 1810
March 5, 1991
Producer(s): Kathleen Bernhardt, NOVA; George Johnson, National Film Board of Canada
Two hundred years ago whales were plentiful in the waters around Canada's Vancouver Island. By the mid 1960s, the whales had all but disappeared due to hunting. During the 1980s, international regulations were enacted to protect these whales. NOVA travels aboard the John Muri to study Vancouver's whales, and analyze the impact of human intervention on these mammals.

Russian Right Stuff: The Mission (The) 1809
February 28, 1991
Producer(s): David Dugan & Bill Reid, NOVA
NOVA follows a team of Soviet cosmonauts from their training at Star City to their record breaking space walk in 1990.

Russian Right Stuff: The Dark Side of the Moon (The) 1808
February 27, 1991
Producer(s): David Dugan & Bill Reid, NOVA
NOVA explores the Soviet manned space program and their ambitious and ultimately unsuccessful bid to reach the moon.

Russian Right Stuff: The Invisible Spaceman (The) 1807
February 26, 1991
Producer(s): David Dugan & Bill Reid, NOVA
This program is the first of three that examines the newly revealed story of the Soviet space program. NOVA follows the origins of the the Soviet space program, from the early liquid-propelled rockets to the first manned space flight, and focuses on Sergei Korolov, the anonymous "chief designer" who guided Soviet aeronautics.

T. rex Exposed 1806
February 19, 1991
Producer(s): Mark Davis, NOVA
Tyrannosaurus rex, the terrifying king of the dinosaurs, recently turned up as a nearly complete skeleton in Montana. NOVA follows the efforts to extract the bones and examines the science and lore of dinosaurs.

Case of the Flying Dinosaur, The 1805
February 12, 1991
Producer(s): Mark J. Davis, NOVA
Are dinosaurs still among us? NOVA looks at the contentious question of whether present-day birds are dinosaurs. Over the years new fossil discoveries keep amending the answer.

Hunt for China's Dinosaurs, The 1804
February 5, 1991
Producer(s): Alain Jehlen, NOVA
NOVA covers the most elaborate expedition ever undertaken in the search for dinosaurs—to China's Gobi Desert. Paleontologists brave sandstorms, heat and worse to find their fossils.

Return to Mt. St. Helens 1801
January 8, 1991
Producer(s): Bebe Nixon, NOVA; John Lynch, BBC
On May 18, 1980, Mt. St. Helens, a long dormant volcano in Washington state, erupted, ravaging 150,000 acres of wilderness. This program reveals how scientists have used Mt. St. Helens as a laboratory to study the ways volcanic eruptions change the environment and create new opportunities for life to return to the mountain.

What's Killing the Children? 1720
December 18, 1990
Producer(s): Larry Klein, Mark Olshaker, Daniel Bailes, NOVA
NOVA tracks a mysterious disease that suddenly and fatally attacks

the children of a small Brazilian town. Researchers from the Centers for Disease Control in Atlanta are called in to crack the case.

In the Land of the Llamas 1719
December 4, 1990
Producer(s): Wolfgang Bayer & Bebe Nixon, NOVA
NOVA profiles the llama, alpaca, vicuna and guanaco of South America. At one time nearly extinct, these four members of the camel family are exceptionally well adapted to life in the beautiful high Andes.

We Know Where You Live 1718
November 27, 1990
Producer(s): Mike Tomlinson, NOVA
Ever wonder how junk mail finds you? NOVA investigates the hidden world of direct marketing, pointing up how advertisers know a lot more about us than we think.

Can the Elephant Be Saved? 1717
November 20, 1990
Producer(s): Noel Buckner & Rob Whittlesey, NOVA
Is the ivory ban in the elephant's best interest? NOVA looks at controversial strategies to save the world's largest land animal from extinction.

Killing Machines 1716
November 13, 1990
Producer(s): Mitchell Koss, NOVA
Robotic weapons that seek out and destroy ships, planes and other targets are the wave of the future. NOVA questions whether their proliferation may spell an end to superpower invincibility.

Earthquake! 1715
November 6, 1990
Producer(s): Carl Charlson, NOVA
NOVA looks at the high-stakes quest to predict earthquakes. Despite

past disappointments, geologists still hope to divine the clues that precede nature's ultimate upheavals.

Blimp Is Back! (The) 1714
October 30, 1990
Producer(s): Michael Barnes, NOVA
NOVA examines the troubled past and promising future of blimps, zeppelins, cyclocranes and other species of airships. There's life in the old gasbags yet.

Poisoned Winds of War, The 1713
October 23, 1990
Producer(s): Michael Rossiter, Central Independent Television
Sixty-five years after attempts to ban them, chemical weapons pose more of a threat than ever. NOVA looks at the problem of controlling substances that are easily produced and cruelly effective.

To Boldly Go... 1712
October 16, 1990
Producer(s): Alain Jehlen, NOVA; Judith Bunting, BBC
NOVA chronicles the Voyager space mission—from Earth to the ends of the solar system. Jupiter, Saturn, Uranus and dozens of moons star in the epic voyage of exploration. Actor Patrick Stewart hosts and narrates.

Neptune's Cold Fury 1711
October 9, 1990
Producer(s): Alain Jehlen, NOVA; Fisher Dilke, BBC
NOVA visits Neptune, the planet that took 12 years for Voyager to reach. Mysteries abound in and around this big, blue world at the outer limits of the solar system. Actor Patrick Stewart hosts and narrates.

KGB, the Computer and Me (The) 1710
October 2, 1990
Producer(s): Robin Bates, NOVA
What happens when a onetime Berkeley hippie/astrophysicist turns detective and gets mixed up with the CIA and the KGB? NOVA follows computer sleuth Clifford Stoll as he tracks down a hacker and thief through a maze of military and research computers.

Will the Dragon Rise Again?: The Genius That Was China 1709
April 3, 1990
Producer(s): Thomas Levenson, NOVA; John Merson & David Roberts, Film Australia
NOVA covers China's long road to economic and technological equality with the West, punctuated by frequent setbacks such as the 1989 massacre of pro-democracy demonstrators in Beijing.

Threat From Japan: The Genius That Was China 1708
March 27, 1990
Producer(s): Thomas Levenson, NOVA; John Merson & David Roberts, Film Australia
East and West came into direct conflict over trade and power in the 19th century resulting in far-reaching changes to both. NOVA explores how Japan was later able to master Western methods while China was not.

Empires in Collision: The Genius That Was China 1707
March 20, 1990
Producer(s): Thomas Levenson, NOVA; John Merson & David Roberts, Film Australia
NOVA examines the extraordinary transformation that propelled Europe outward into the world from the 15th to 18th centuries, while China remained the insular middle kingdom

Rise of the Dragon: The Genius That Was China 1706
March 13, 1990
Producer(s): Thomas Levenson, NOVA; John Merson & David Roberts, Film Australia
13th century China was the richest, most powerful, most technologically advanced civilization on earth. NOVA looks at what China achieved and how Chinese politics, culture and economy kept it from doing more.

Big Spill, The 1705
February 27, 1990
Producer(s): Denise DiIanni, NOVA
Covering the 1989 Exxon Valdez oil spill from an unexplored angle, NOVA focuses on how technology failed in preventing, containing and cleaning up the Alaskan disaster.

Bomb's Lethal Legacy, The 1704
February 13, 1990
Producer(s): Noel Buckner & Rob Whittlesey, NOVA
NOVA examines an alarming nuclear waste problem at the Hanford Nuclear Reservation in eastern Washington state, where 45 years of mismanagement in the nuclear weapons industry will cost billions to correct.

Disguises of War 1703
February 6, 1990
Producer(s): David Barlow & John Alexanders, Channel 4
NOVA reveals the art of deception in war—from simple camouflage to the expensive, radar-evading technology embodied in the B-2 Stealth bomber.

Race for the Top 1702
January 23, 1990
Producer(s): Joe Blatt, NOVA
Using some of the largest machines ever built, American and Euro-

pean physicists race to discover one of the most fundamental and most elusive objects in nature—the top quark.

Poison in the Rockies 1701
January 9, 1990
Producer(s): Denise DiIanni, NOVA; Christopher McLeod & Robert Lewis, EIF
NOVA reports on the 100 year old legacy of pollution from mining that poisons the once-pristine waters of the Rocky Mountain states. Acid rain and economic development also contribute to stress on the West's scarce water supply

Schoolboys Who Cracked the Soviet Secret (The) 1620
December 12, 1989
Producer(s): Denise DiIanni, NOVA; Neville Bolt, BBC
NOVA re-enacts a classic case of classroom detection when English schoolboys track down a secret Soviet launch site.

Yellowstone's Burning Question 1619
December 5, 1989
Producer(s): Kathleen Bernhardt & Ray Paunovich, NOVA
The 1988 Yellowstone fire may have been one of the worst in human memory, but nature has had eons of experience with such events. NOVA accompanies scientists who are studying the surprisingly rapid recovery from the blaze.

What Is Music? 1618
November 21, 1989
Producer(s): John Angier, NOVA
NOVA explores the science of musical sound—from what makes a classic violin to how the human brain perceives music. Bells, trumpets, human voices and computers all perform.

Magic Way of Going: The Story of Thoroughbreds (A) 1014
November 15, 1989
Producer(s): Bebe Nixon, NOVA; John Basset, CBC
Can the thoroughbred horse run any faster? NOVA examines the billion-dollar horse racing industry in its search for the magic combination of speed, stamina and the will to win.

Will Venice Survive Its Rescue? 1617
November 14, 1989
Producer(s): Nancy Porter, NOVA
Increasingly awash in high water, the romantic city of Venice is counting on high-tech floodgates to save it from drowning. Environmentalists worry that the gates may destroy the fragile lagoon that surrounds the city.

Hurricane! 1616
November 7, 1989
Producer(s): Larry Engel & Tom Lucas, BBC
NOVA studies hurricanes—the lurking giants waiting to destroy many coastal areas—by flying straight into one. Scientists hope that such close-up studies will supply the data to make better predictions.

Decoding the Book of Life 1615
October 31, 1989
Producer(s): Jon Palfreman, BBC
Biologists around the world gear up to decode the three-billion-letter genetic message that describes how humans are made. Ethicists warn that it may not be such a good idea.

Echoes of War 1614
October 24, 1989
Producer(s): Linda Garmon, NOVA
The atomic bomb might have ended World War II, but radar was the quiet miracle that won battles. NOVA tells the little-known wartime history of radar.

Design Wars! 1613
October 17, 1989
Producer(s): Marian Marzynski, NOVA
Five architects compete for the approval of architecture-obsessed Chicagoans in the contest to build the city's new public library. NOVA looks at the strengths and weakness of each of the surprisingly varied entries.

Controversial Dr. Koop (The) 1612
October 10, 1989
Producer(s): Susanne Simpson, NOVA
In this profile of the former Surgeon General, NOVA follows events as they unfold in a unique behind-the-scenes account of a man who speaks his mind on AIDS, smoking, and abortion.

Hidden City (The) 1611
October 3, 1989
Producer(s): Carl Charlson, NOVA
Actor Judd Hirsch narrates this behind-the scenes look at what makes New York City tick. Water, power and waste are the critical systems that an urban population depends upon working normally, but when they break down, the consequences can be disastrous . [17th season premiere]

Confronting the Killer Gene 1610
March 28, 1989
Producer(s): Alain Jehlen, NOVA; Theresa Hunt & Jack Weber, BBC
Arlo, Nancy and Janice each have a 50/50 chance of developing a devastating nerve disorder. A new laboratory test can give them the answer. Would you take the test? Thousands of others face a similar choice: whether to find out if they carry the genetic time bomb of Huntington's disease. NOVA looks at this incurable disease which affects 20,00 people in the US and threatens tens of thousands of others.

World is Full of Oil! (The) 1609
March 21, 1989
Producer(s): Andrew Liebman & Gail Ringel, NOVA
Scientific detectives test their ingenuity in the effort to find underground oil deposits.

Legends of Easter Island 1608
March 14, 1989
Producer(s): Bebe Nixon, NOVA; Bettina Lerner & John Lynch, BBC
Ancient legends hold the clues to the violent history of the South Pacific's Easter Island.

Secrets of Easter Island 1607
March 7, 1989
Producer(s): Bebe Nixon, NOVA; Bettina Lerner & John Lynch, BBC
NOVA investigates the mystery of Easter Island in the South Pacific. Who built its celebrated statues and why?

Adrift on the Gulf Stream 1606
February 28, 1989
Producer(s): John Borden, NOVA
NOVA explores the importance of the Gulf Stream to ocean life, climate and human history.

God, Darwin and Dinosaurs 1605
February 21, 1989
Producer(s): Larry Engel & Tom Lucas, NOVA
In an Idaho classroom, teacher Phil Gerrish puts an unorthodox interpretation on the day's biology lesson. As students take notes, he explains that creationism is a valid scientific explanation for the origin on life. Once relying solely on the literal word of the Bible to make their case, creationists now argue that the scientific evidence is

on their side. NOVA reports on this new twist in the long-running battle between creationism and evolution.

Back to Chernobyl 1604
February 14, 1989
Producer(s): Denise DiIanni & Bill Kurtis, NOVA
NOVA goes to the Soviet Union for an inside investigation of the world's most catastrophic nuclear power accident with correspondent Bill Kurtis.

Strange New Science of Chaos (The) 1603
January 31, 1989
Producer(s): Jeremy Taylor, NOVA
NOVA explains "chaos," a new science that is making surprising sense out of chaotic phenomena in nature, from the weather to brain waves.

Last Journey of a Genius (The) 1602
January 24, 1989
Producer(s): Denise DiIanni, NOVA; Christopher Sykes, BBC
NOVA looks at the bongo-playing scientist, adventurer, safecracker and yarn-spinner Richard Feynman, most recently famous for his role as gadfly of the Presidential Commission investigating the explosion of the space shuttle Challenger.

Hot Enough for You? 1601
January 17, 1989
Producer(s): Kathleen Bernhardt, NOVA; Peter Ceresole, BBC
Was the searing summer of 1988 a taste of things to come? NOVA looks at the greenhouse effect, which portends higher temperatures, rising sea levels and other environmental disasters.

Can We Make a Better Doctor? 1521
December 13, 1988
Producer(s): Michael Barnes, NOVA
NOVA embarks on a 10-year project to profile—in its entirety—the education of a doctor. In the premiere episode, NOVA follows a select group of students as they start their freshman year at Harvard Medical School under a revolutionary program emphasizing early clinical contact with patients.

All-American Bear (The) 1520
December 6, 1988
Producer(s): Bebe Nixon & Ray Paunovich, NOVA
The life of the shy, intelligent black bear in the wild—foraging, mating, playing and constantly preparing for its remarkable hibernation—is captured for the first time on film by NOVA.

Light Stuff (The) 1519
November 22, 1988
Producer(s): Mark Davis, NOVA
Reliving a Greek myth takes an effort of mythic proportions. NOVA looks at the human powered-flight across the Aegean Sea, a journey that symbolically recreated the mythical flight of Daedalus. The plane, respectfully named Daedalus 88, is examined from the early prototypes to its dramatic landing in the surf after a 74-mile flight from the island of Crete to Santorini.

Who Shot President Kennedy? 1518
November 15, 1988
Producer(s): Robert Richter, NOVA
Using previously unavailable technology, NOVA probes the available evidence surrounding the 1963 assassination of John F. Kennedy.

Do Scientists Cheat? 1517
October 25, 1988
Producer(s): Noel Buckner & Rob Whittlesey, NOVA
NOVA examines the troubling question of scientific fraud: How prevalent is it? Who commits it? And what happens when the perpetrators are caught?

Can the Vatican Save the Sistine Chapel? 1515
October 14, 1988
Producer(s): Susanne Simpson, NOVA
Science meets art in the controversial effort to restore the frescoes of Michelangelo's famous Sistine Chapel.

Can the Next President Win the Space Race? 1516
October 11, 1988
Producer(s): Geoffrey Haines-Stiles, NOVA
Thirty years after Sputnik, the United States space program is mired in uncertainty, while the Russians, Europeans, Japanese and others sprint onward and upward.

Beyond the Knife: Pioneers of Surgery 1514
September 27, 1988
Producer(s): Jon Palfreman, NOVA & BBC
NOVA examines some of the excesses of surgery and at how new drugs and technologies are rendering some operations obsolete.

New Organs for Old: Pioneers of Surgery 1513
September 20, 1988
Producer(s): Jon Palfreman, NOVA & BBC
NOVA takes a look at perhaps the most remarkable of all surgical procedures—the replacement of a diseased organ with one transplanted from another human. Kidneys, hearts, livers, lungs—all presenting unique problems for the surgeon—are among the spare parts now routinely exchanged from the dead to the dying.

Into the Heart: Pioneers of Surgery 1512
September 13, 1988
Producer(s): Jon Palfreman, NOVA & BBC
Once unthinkable, open-heart surgery is now an everyday miracle. NOVA recounts the daring attempts to operate on the most forbidden organ of all: the heart. The contemporary view of the surgeon-as-a-miracle-worker owes largely to those surgical successes, realized during two decades beginning in the 1940s, dealing with this crucial pump which was long considered too fragile to fix.

Brutal Craft: Pioneers of Surgery (The) 1511
September 6, 1988
Producer(s): Fionna Holmes & Jon Palfreman, NOVA & BBC
Part one of a four-part series on the pioneers of modern surgery. NOVA takes a look at the early days of surgery and the gradual development first of anesthesia, then antiseptics and later blood transfusions—three revolutions that made modern surgery possible. [16th season premiere]

Can You Still Get Polio? 1510
April 5, 1988
Producer(s): Nancy Porter, NOVA
Most cases of polio in this country are caused by the vaccine designed to prevent it. NOVA examines the controversy surrounding the nation's vaccine policy.

Race for the Superconductor 1509
March 29, 1988
Producer(s): Linda Garmon, NOVA
NOVA charts an electronics revolution in the making as Japan and the United States race to develop a material that will conduct electricity at room temperature with zero resistance.

Man Who Loved Numbers (The) 1508
March 22, 1988
Producer(s): Alain Jehlen, NOVA; Christopher Sykes, Channel 4
NOVA explores the life of Srinivasa Ramanujan, a poor clerk from India who astounded mathematicians in the 1910s with his brilliant insight into the world of numbers.

Mystery of the Master Builders 1507
March 15, 1988
Producer(s): Robin Bates, NOVA
Princeton professor and author Robert Mark tracks down the engineering secrets of some of the beautiful buildings in the world including Notre Dame in Paris, St. Paul in London and the Roman Pantheon.

Whale Rescue 1507.1
March 15, 1988
Producer(s): Mark Davis, NOVA
Three baby whales are saved from a Cape Cod beaching, nursed back to health and then returned to the wild in the first-ever rescue and release of the giant sea animals.

Battles in the War on Cancer: Breast Cancer—Turning the Tide 1506
March 1, 1988
Producer(s): Graham Chedd & Alice Markowitz, NOVA
Breast cancer claims the lives of four American women every hour. Jane Pauley of NBC News hosts and narrates this NOVA report on stepped-up efforts to reduce the death rate from this all-too-common killer.

Battles in the War on Cancer: A Wonder Drug on Trial 1505
February 23, 1988
Producer(s): Graham Chedd & Alice Markowitz, NOVA
In part one of a two-part special presentation, NOVA reports on the trials to determine whether the new drug Interleukin-2—the first to make use of the body's own disease-fighting strategy—will live up to

its promise as a pivotal cancer breakthrough. Jane Pauley of NBC News hosts and narrates.

Why Planes Burn 1504
February 9, 1988
Producer(s): Bebe Nixon, NOVA; Alec Nisbett, BBC
Airplane fires are often deadly. NOVA looks at efforts to make fires aboard planes less likely and more survivable.

Buried in Ice 1503
February 2, 1988
Producer(s): Denise DiIanni, NOVA ; Nicholas Bakyta, Tinsel Media Productions
Scientists investigate the frozen remains of members of the 19th century Franklin Expedition to the Canadian Arctic and ask why all perished.

How to Create a Junk Food 1502
January 26, 1988
Producer(s): Kathleen Bernhardt, NOVA; Edward Poulter & Mike Cockburn, Channel 4
Julia Child introduces NOVA's behind-the-scenes look at how science aids in the creation of snack foods.

Secrets of the Lost Red Paint People 1420
December 15, 1987
Producer(s): Bebe Nixon, NOVA; T.W. Timreck
NOVA follows archeologists as they unearth clues, some 7,000 years old, about an unknown, mysterious and advanced sea faring people who lived along the North Atlantic coast of the United States and Canada.

Riddle of the Joints 1419
December 8, 1987
Producer(s): Denise DiIanni, NOVA; Katherine Everett, BBC
A trail of evidence leading from a medieval abbey to a small town in

Connecticut sheds new light on rheumatoid arthritis, a crippling inflammation of the joints with no known cause or cure.

Ancient Treasures from the Deep 1418
December 1, 1987
Producer(s): Denise DiIanni, NOVA; Jack Kelley, KUHT TV-Houston
NOVA joins underwater archeologists as they explore the oldest shipwreck ever excavated, a richly-laden merchant vessel dating from the time of King Tut.

How Good is Soviet Science? 1417
November 17, 1987
Producer(s): Martin Smith, NOVA
NOVA takes a behind-the-scenes look at science and technology in the USSR, where the government is trying novel approaches in the effort to catch up with the West.

Volcano! 1416
November 10, 1987
Producer(s): Alain Jehlen, NOVA; John Simmons, BBC
Millions live in the shadows of nature's ticking time-bombs—volcanos. NOVA accompanies scientists who are developing new techniques to predict when and how explosively volcanos will erupt.

A Man, A Plan, A Canal, Panama 1415
November 3, 1987
Producer(s): Carl Charlson, NOVA
The Panama Canal opened in 1914 after a 30-year effort that dwarfed the building of the pyramids. Historian David McCullough navigates through the canal and tells the story of the human drama behind the engineering feat.

Japan's American Genius 1414
October 27, 1987
Producer(s): Marian Marzynski, NOVA
Is Detroit inventor Stanford Ovshinsky the new Thomas Edison? Japanese industries are betting that the genius behind the amorphous materials is onto something big.

Hidden Power of Plants 1413
October 20, 1987
Producer(s): Kathleen Bernhardt, NOVA; Vishnu Mathur, CBC
Plants produce some of the world's most potent chemicals in the fight against disease. NOVA follows the urgent efforts to track down new medicines in nature.

Spy Machines 1412
October 13, 1987
Producer(s): Lew Allison & Blaine Baggett, NOVA
On the 25th anniversary of the Cuban missile crisis, NOVA investigates the spy planes and satellites that played a critical role in that chapter in history and influenced arms control today.

Death of a Star 1411
October 6, 1987
Producer(s): Robin Bates, NOVA
A star blows itself apart in a nearby galaxy, and astronomers scramble to study the rare event. NOVA covers a fast-breaking science story as it is happening. [15th season premiere]

Rocky Road to Jupiter 1410
April 7, 1987
Producer(s): Geoffrey Haines-Stiles, NOVA
In a case study of the strengths and weaknesses of the United States space program, NOVA chronicles the ambitious and long-delayed

Galileo mission to Jupiter—still on the ground long after its planned May 1986 launch.

Desert Doesn't Bloom Here Anymore (The) 1409
March 31, 1987
Producer(s): Kathleen Bernhardt, NOVA; Mike Andrews, BBC
In rich and poor countries alike, once-productive farms are turning to desert because of mismanagement of water resources. NOVA examines the causes and cures of desertification.

Will the World Starve? 1408
March 24, 1987
Producer(s): Kathleen Bernhardt, NOVA; Mike Andrews, BBC
All over the world, farmers are taking more from the soil than they return. NOVA reports on the soil crisis in world agriculture—a plight that already has resulted in massive starvation.

Great Moments from NOVA 1407.1
March 10, 1987
Producer(s): Betsy Anderson, NOVA
NOVA presents two hours of the best from its 14 seasons of exciting science coverage. A "talking" chimp, an exploding volcano and a sight-and-sound space video are but a few of the memorable segments. Richard Kiley hosts.

Confessions of a Weaponeer 1407
March 3, 1987
Producer(s): Alain Jehlen, NOVA; Ann Druyan, Carl Sagan Productions
Harvard chemist George Kistiakowsky was an anti-Bolshevik soldier in 1919 Russia, an atomic bomb scientist at Los Alamos, a presidential advisor in the Eisenhower White House and an arms control ac-

tivist. Shortly before his death he recounts his eventful career to interviewer Carl Sagan.

Hole in the Sky (The) 1406
February 24, 1987
Producer(s): Linda Harrar, NOVA
NOVA travels to Antarctica with an emergency scientific expedition to study a baffling "hole" in the Earth's protective ozone layer.

Freud Under Analysis 1405
February 17, 1987
Producer(s): Susanne Simpson, NOVA
Fifty years after his death, the creator of psychoanalysis is still the subject of intense debate. Was Freud right or wrong? NOVA profiles the enigmatic man and his controversial legacy.

Orangutans of the Rain Forest 1404
February 10, 1987
Producer(s): Wolfgang Bayer, NOVA
NOVA cameras travel to Borneo, one of the last habitats of the wild orangutans, where scientists study the endangered ape. Who is observing whom? It is not always clear.

Why Planes Crash 1403
February 3, 1987
Producer(s): Veronica Young, NOVA
Between 60 and 80 percent of all commercial airplane accidents are attributable to pilot error. NOVA looks at some shocking instances of pilot negligence and what airlines are doing to solve the problem.

Children of Eve 1402
January 27, 1987
Producer(s): Denise DiIanni, NOVA; John Groom, BBC
NOVA examines a controversial theory that traces our ancestry to a small group of women living in Africa 300,000 years ago.

Countdown to the Invisible Universe 1401
January 20, 1987
Producer(s): Kathleen Bernhardt, NOVA; Glyn Jones, Quanta Limited
NOVA scans the universe with the infrared eye of IRAS—the infrared Astronomical Satellite—and discovers never-before-seen comets, stars, galaxies and other celestial wonders and enigmas.

Top Gun and Beyond 1501
January 19, 1987
Producer(s): Chris Haws, NOVA
Today's sophisticated fighter jets can almost fly themselves, but well-trained pilots are still needed to win air battles. NOVA looks at how planes and pilots are adapting to high technology.

How Babies Get Made 1320
January 13, 1987
Producer(s): Barbara Costa, NOVA; Jon Palfreman, BBC
NOVA explores the ground-breaking experiments that led to the discovery of a tiny sequence of molecules—and more clues to the mystery of how a complete baby develops from a single cell.

Leprosy Can Be Cured! 1319
December 16, 1986
Producer(s): Michael Barnes, NOVA
Leprosy, a misunderstood disease that has been curable for 40 years, still afflicts some 12 million people. NOVA looks at the tragedy of the disease that need not be.

Sail Wars! 1318
December 9, 1986
Producer(s): Noel Buckner & Rob Whittlesey, NOVA
Yankee ingenuity has designs on the America's Cup. NOVA goes behind-the-scenes to look at the engineering effort to design a technically advanced sailboat.

Are You Swimming in a Sewer? 1317
December 2, 1986
Producer(s): Noel Buckner & Rob Whittlesey, NOVA
NOVA dips into the sad plight of our coastal waters, where toxic chemicals, raw sewage and disease-carrying microbes are routinely dumped.

Mystery of the Animal Pathfinders 1316
November 25, 1986
Producer(s): Neil Goodwin, NOVA
Birds do it; bees do it, butterflies, bats and eels do it—all leave one habitat to migrate to another, often thousands of miles away. NOVA penetrates the mystery of where animals migrate, why and how they get there.

Is Anybody Out There? 1315
November 18, 1986
Producer(s): Geoffrey Haines-Stiles, NOVA
Could there be life beyond Earth? Only now is it possible to scan the skies in a systematic attempt to find out. NOVA joins the search with guest host Lily Tomlin.

Can AIDS Be Stopped? 1314
November 11, 1986
Producer(s): Kathleen Bernhardt, NOVA; David Dugan & Max Whitby, BBC
What are the prospects for halting or curing the deadliest epidemic ever to challenge modern medicine? NOVA finds cause for both hope and alarm in the battle against AIDS.

High-Tech Babies 1313
November 4, 1986
Producer(s): Nancy Porter, NOVA
Scientific breakthroughs now make it possible to reproduce ourselves in ways never before imagined. NOVA looks at the medical, legal and moral questions raised by this brave new technology.

Planet that Got Knocked on its Side (The) 1312
October 21, 1986
Producer(s): Bebe Nixon, NOVA; Fisher Dilke, BBC
The adventures of the Voyager 2 spacecraft continues as it passes the rings of Uranus. Scientists suspect that violent events in the early history of the planet may have shaped Uranus and its strange collection of moons.

Search for the Disappeared (The) 1311
October 14, 1986
Producer(s): David Dugan, NOVA
NOVA joins scientists in Argentina as they help locate kidnapped children and identify thousands of dead in the aftermath of a military reign of terror. [14th season premiere]

Visions of Star Wars 1310.1
April 22, 1986
Producer(s): Graham Chedd & Andrew Liebman, NOVA; Janet McFadden, FRONTLINE
NOVA and Frontline combine resources to explore the Strategic Defense Initiative. The two-hour documentary contains the most comprehensive information on "Star Wars" ever produced. Bill Kurtis of WBBM-TV/Chicago hosts.

When Wonder Drugs Don't Work 1310
March 25, 1986
Producer(s): Salem Mekuria, NOVA; David Dugan, BBC
NOVA examines the medical community's alarm as the spread of antibiotic resistant infection increases, and studies how one hospital fights its own dramatic epidemic.

Rise of a Wonder Drug (The) 1309
March 18, 1986
Producer(s): Kathleen Bernhardt, NOVA; David Dugan, BBC
When Alexander Fleming discovered the penicillin mold in 1928, he

never considered its possible therapeutic value. NOVA explores the "Fleming myth" and reveals the true story of the men who worked behind the scenes to develop the wonder drug of the century.

Return of the Osprey 1308
March 11, 1986
Producer(s): Kathy White, NOVA; Judy Fieth & Michael Male, Blue Earth Films
NOVA follows a conservation success story as environmentalists, scientists and bird-lovers fight to save the majestic Osprey from extinction.

Skydive to the Rain Forest 1307
March 4, 1986
Producer(s): Salem Mekuria, NOVA; Adrian Warren, BBC
NOVA journeys to a remote region of southern Venezuela where the land is alive with spectacular waterfalls, colored by exotic flowers and inhabited by rare species of birds and animals.

Toxic Trials 1306
February 25, 1986
Producer(s): John Angier, NOVA
When a high number of cancer cases struck the suburban community of Woburn, Massachusetts, the town mobilized to investigate why. The result was a landmark study of the effects of hazardous wastes. NOVA explores the legal and scientific implications of the link between environmental pollution and illness.

Case of the Frozen Addict (The) 1305
February 18, 1986
Producer(s): John Palfreman, NOVA
In July 1982, a 42-year-old addict in a San Jose, California jail became paralyzed—unable to more or talk. His symptoms, caused by a bad batch of synthetic heroin, were indistinguishable from those associated with Parkinson's disease, a degenerative nerve disorder that

strikes the elderly. NOVA traces the story of a "designer" drug which could lead to a major medical breakthrough.

Life's First Feelings 1304
February 11, 1986
Producer(s): James Lipscomb & Bill Wander, NOVA
NOVA explores the incredibly complex emotional development of infants and examines the current theory that early childhood psychological intervention can head off emotional problems later in life.

Horsemen of China (The) 1303
February 4, 1986
Producer(s): Susanne Simpson, NOVA; Andre Singer, Granada TV
For centuries, the Chinese Kazakh horsemen preserved their ancient traditions, refusing to be dominated by either the Chinese or nearby Russian cultures. Today, however, this nomadic tribe has integrated communism into its way of life. NOVA traces the ancient Kazakh lifestyle and looks at how the Chinese cultural Revolution has modernized Kazakh customs.

Goddess of the Earth 1302
January 28, 1986
Producer(s): Barbara Costa, NOVA; John Groom, BBC
Gaia, the Greek word for Earth goddess, also is the name of the controversial hypothesis that life on Earth controls the environment. NOVA explores this provocative theory that challenges conventional ways of thinking about the Earth.

Halley's Comet: Once in a Lifetime 1301
January 21, 1986
Producer(s): Linda Garmon, NOVA; Alec Nisbett, BBC
NOVA observes worldwide preparations as amateur comet hunters, astronomers and scientists armed with specialized cameras, high pow-

ered telescopes and spacecraft look to the heavens in search of the expected arrival in 1986 of Halley's Comet.

Plane that Changed the World (The) 1220
December 17, 1985
Producer(s): Marty Ostrow, NOVA
NOVA joins the 50th anniversary celebration of the DC-3—the plane that revolutionized commercial air travel, served gallantly in World War II and is called the most important plane ever built.

Animal Architects 1219
December 3, 1985
Producer(s): Denise DiIanni, NOVA; Malcolm Penny, Oxford Scientific Films
NOVA examines the intricate world of nature's construction industry and presents rare footage of unusual habitats.

Genetic Gamble (The) 1218
November 26, 1985
Producer(s): Stuart Harris, NOVA
NOVA examines current research and its ethical implications as modern medicine confronts the era of human gene therapy.

Tornado! 1217
November 19, 1985
Producer(s): Larry Engel & Tom Lucas, NOVA
NOVA follows a chase team—a group of scientists who chart deadly tornadoes—in an effort to learn more about predicting nature's most powerful and elusive weather phenomenon.

Child Survival: The Silent Emergency 1216
November 12, 1985
Producer(s): Linda Harrar, NOVA
NOVA charts the progress of an ambitious worldwide health program

established to save the lives of millions of children who continue to die from common but curable diseases.

Magic of Special Effects (The) 1215
November 5, 1985
Producer(s): Bebe Nixon, NOVA; Edward Goldwyn, BBC
NOVA cameras go behind-the-scenes to reveal the new art of illusion, Hollywood-style, focusing on three blockbuster films—"Return of the Jedi," "Indiana Jones and the Temple of Doom" and "2010: The Year We Made Contact."

Robot Revolution? (The) 1214
October 29, 1985
Producer(s): Brian Kaufman, NOVA
How are the computer and the robot affecting the way we work? NOVA chronicles the new industrial revolution reshaping the American workplace.

What Einstein Never Knew 1213
October 22, 1985
Producer(s): Melanie Wallace, NOVA; James Burge & Andrew Millington, BBC
NOVA follows the quest of a new generation of physicists to do what Albert Einstein could not—find a single law to explain all physical phenomena. In a remarkable experiment Carlo Rubbia and Simon van de Meer discovered proof of the link between two of the four known forces: "electromagnetic" and "weak". This film examines the impact of this breakthrough.

Seeds of Tomorrow 1212
October 15, 1985
Producer(s): Graham Chedd, NOVA
NOVA examines worldwide efforts of scientists who employ aggressive agricultural technologies to ensure food for the future.

National Science Test II (The) 1211
October 8, 1985
Producer(s): Theodore Bogosian, NOVA
In NOVA's special sequel to1984's National Science Test, viewers can match wits with celebrity panelists David Attenborough, Michelle Johnson, Edwin Newman and Alvin Poussaint and a live studio audience. Art Fleming hosts. [13th season premiere]

Monarch of the Mountains 1210
March 19, 1985
Producer(s): Bebe Nixon, NOVA; Ray Paunovich, Mountain West Films
NOVA explores the breeding, migration and survival patterns of the Rocky Mountain elk in a unique film, made totally under natural conditions. Telephoto lenses were used so as not to disturb the animals; filmmakers spent 18 months tracking the elk through the breathtaking Wyoming Rockies.

Child's Play: Prodigies and Possibilities 1209
March 12, 1985
Producer(s): Noel Buckner, Janet Mendelsohn & Rob Whittlesey, NOVA
What do Wolfgang Amadeus Mozart, the painter Raphael and chess champion Bobby Fischer have in common? They were all child prodigies. NOVA explores the current efforts to learn more about the nature of giftedness.

Mathematical Mystery Tour (The) 1208
March 5, 1985
Producer(s): Susanne Simpson, NOVA; Jon Palfreman, BBC
Imagine a bottle with no inside or a number bigger than infinity or parallel lines that meet. Welcome to the world of pure mathematics. NOVA offers a look into the wholly abstract, quirky world of mathematics.

Baby Talk 1207

February 26, 1985

Producer(s): Melanie Wallace, NOVA; Suzanne Campbell-Jones, BBC

How children acquire language is a mystery. Does the process begin in the womb? And which comes first, language or thought? NOVA explores the fascinating world of baby talk and reveals the latest theories on this remarkable achievement.

Shape of Things (The) 1206

February 19, 1985

Producer(s): Susanne Simpson, NOVA; Neil Goodwin, Peace River Films

Sea shells, crystals, honeycombs, eggs and seeds: They are shaped the way they are for a reason. NOVA takes viewers on a unique journey of discovery to find out why things are shaped the way they are and why they work so well.

AIDS: Chapter One 1205

February 12, 1985

Producer(s): Thea Chalow, NOVA

Acquired Immune Deficiency Syndrome, or AIDS, is a deadly disease that has struck down some 2,000 people in the four years since its discovery. NOVA examines how modern science has been unraveling the mystery of this baffling ailment.

In the Land of the Polar Bears 1204

February 5, 1985

Producer(s): Denise DiIanni, NOVA; Yuri Ledin, Norilsk TV Studio, USSR

A rare look at the beautiful and desolate Wrangel Island—a Soviet possession 300 miles off the coast of Alaska—as seen through the eyes of Soviet Filmmaker and naturalist Yuri Ledin. Wrangel Island is not only the home to Siberian snow geese, polar foxes and walruses, but serves as the world's largest denning area for polar bears.

Conquest of the Parasites 1203
January 29, 1985
Producer(s): Terry Kay Rockefeller, NOVA; Jon Palfreman, BBC
NOVA examines the complex world of parasites, parasitic diseases and the exciting work currently being done by a new breed of medical researchers as they meet the challenge of conquering the world's number one medical problem.

Edgerton and His Incredible Seeing Machines 1201
January 15, 1985
Producer(s): Bebe Nixon, NOVA; Gary Hochman, Nebraska ETV Network
NOVA explores the fascinating world of Dr. Harold Edgerton, electronics wizard and inventor extraordinaire, whose invention of the electronic strobe, a "magic lamp," has enabled the human eye to see the unseen.

Garden of Inheritance (The) 1119
January 8, 1985
Producer(s): Peter Crawford, BBC
In this docudrama presentation, NOVA looks at the life, times and work of Gregor Mendel, the 19th century Augustinian friar whose revolutionary scientific experiments in selective breeding have made him the "Father of Genetics."

Global Village 1202
January 2, 1985
Producer(s): Richard Keefe, BBC
NOVA presents an in-depth look at India's attempt to use satellite technology to leapfrog into the era of space-age communication and whether it brings benefit or blight to India's villages and rural areas.

Eyes Over China 1020
December 27, 1984
Producer(s): Brian Kaufman, NOVA
NOVA documents a dramatic encounter in international medicine when an American plane lands in China—equipped with a state-of-the-art eye-operating theater—and two very different medical systems meet eyeball to eyeball.

Stephen Jay Gould: This View of Life 1118
December 18, 1984
Producer(s): Linda Harrar, NOVA
What do dinosaurs, a panda's thumb and a peacock's tail have in common? Dr. Stephen Jay Gould, the internationally renowned paleontologist and evolutionary theorist, provides some surprising answers in this NOVA profile.

Acid Rain: New Bad News 1117
December 11, 1984
Producer(s): John Angier, NOVA
The debate over "acid rain" continues to grow. NOVA travels to West Germany, the mid-Atlantic states and New England to examine the controversy surrounding this phenomenon.

Space Women 1115
November 27, 1984
Producer(s): Melanie Wallace, NOVA
They have been part of the United States' space program for more than 20 years. Who are these talented, courageous women? NOVA looks at astronaut Sally Ride and her colleagues, how they are trained and their role in NASA's future.

Jaws: The True Story 1116
November 27, 1984
Producer(s): Melanie Wallace, NOVA; Tony Salmon, BBC
Acclaimed underwater cameraman Al Giddings takes NOVA view-

ers beneath the waves to explore the fact and fiction surrounding the great white shark.

Frontiers of Plastic Surgery 1114
November 20, 1984
Producer(s): Theodore Bogosian, NOVA
NOVA's sequel to "A Normal Face" examines the merging of technology and art in modern reconstruction and cosmetic surgical techniques.

Farmers of the Sea 1113
November 13, 1984
Producer(s): Melanie Wallace, NOVA; Jim Larison, Oregon State University
NOVA looks at the "blue revolution"—modern advances in the ancient art of raising aquatic animals and plants—in the United States, Japan, Scotland and other countries.

Nomads of the Rain Forest (The) 1112
November 6, 1984
Producer(s): Grant Behrman & Melanie Wallace, NOVA
NOVA visits a tribe of Ecuadoran natives who still maintain traditions that date back to the Stone Age—thirty years after their first contact with Western Civilization.

Mystery of Yellow Rain (The) 1111
October 30, 1984
Producer(s): Susanne Simpson, NOVA ; Jeremy Taylor, BBC
NOVA explores whether "yellow rain," described by members of the Hmong tribe of Laos, is a form of chemical warfare—or a naturally occurring phenomenon.

Fountains of Paradise (The) 1110
October 20, 1984
Producer(s): Melanie Wallace, NOVA; Sandra Nichols
NOVA explores the billion-dollar-plus Mahaweli Irrigation Project

in Sri Lanka. Will this high-risk project prove to be a great leap forward or an industrial and sociological disaster?

National Science Test I (The) 1109
October 16, 1984
Producer(s): Theodore Bogosian, NOVA
NOVA departs from tradition with the first National Science Test. Viewers can match wits with celebrity panelists Jane Alexander, Jules Bergman, Marva Collins and Edwin Newman. Art Fleming hosts. [12th season premiere]

Space Bridge to Moscow 1120
October 2, 1984
Producer(s): Bebe Nixon & Terry Kay Rockefeller, NOVA
At a time when scientific exchange between the United States and the Soviet Union is at its lowest since the 1950s, a special hookup will allow eight leading Soviet and American scientists to share ideas face-to-face before millions of television viewers in each country on this NOVA special.

World According to Weisskopf (The) 1108
April 3, 1984
Producer(s): Brian Kaufman, NOVA
Victor Weisskopf: physicist, lover of music and citizen of the world. NOVA profiles the international statesman of science and learns that one of the giants of 20th century physics is also one of the country's greatest humanists.

Make My People Live: The Crisis in Indian Health Care 1107
March 27, 1984
Producer(s): Linda Harrar, NOVA
What are America's obligations to its native population? As an important Indian health act comes up for renewal in Congress in the spring of 1984, NOVA explores the state of medical care for a proud but vulnerable minority.

Down on the Farm 1106
March 20, 1984
Producer(s): Noel Buckner & Rob Whittlesey, NOVA
Agriculture is America's biggest industry. This productivity, envied around the world, is also depleting the most essential ingredients in farming: water and soil. NOVA looks at the agricultural dilemma, the short term need for profit and long term needs of the land.

Visions of the Deep: The Underwater World of Al Giddings 1105
March 6, 1984
Producer(s): Melanie Wallace, NOVA; Tony Salmon, BBC
Al Giddings is one of the greatest underwater photographers in the world. In a riveting look at the unearthly beauties and terrors of the seas, NOVA presents a portrait of Giddings at work.

Will I Walk Again? 1104
February 28, 1984
Producer(s): Melanie Wallace, NOVA; Graham Hurley, Channel 4
Is there a cure for paralyzing spinal injuries? Most neurosurgeons are doubtful, pointing to the central nervous system's most apparent inability to heal itself. But others dispute the point. NOVA explores the debate, the hopes for a cure and recent breakthroughs to help paralyzed patients.

China's Only Child 1103
February 14, 1984
Producer(s): Terry Kay Rockefeller, NOVA ; Edward Goldwyn, BBC
Efforts to control the population explosion are among the burning controversies of our time. NOVA looks at the one-child policy of the People's Republic of China, a revolutionary decree with profound implications for a people accustomed to traditionally large families.

Antarctica: Earth's Last Frontier 1102
January 31, 1984
Producer(s): Bebe Nixon, NOVA; Tom Williamson, New Zealand National Film Unit
An astronaut once observed a great white light shining out from the bottom of our world: Antarctica, the ice-covered continent we are only just beginning to understand. NOVA visits this wilderness of ice, larger than the United States and Mexico combined, whose only warm-blooded residents are seals, skuas, penguins and scientists.

Case of ESP (The) 1101
January 17, 1984
Producer(s): Terry Rockefeller, NOVA; Tony Edwards, BBC
In the past decade, a number of researchers have begun systematic laboratory research into extrasensory perception—ESP. NOVA considers the claims for—and against—paranormal phenomena and looks at some startling applications in the field of archaeology, criminology and warfare.

Alcoholism: Life Under The Influence 1021
January 10, 1984
Producer(s): Thea Chalow, NOVA
In a culture laced with alcohol, the search for a scientific understanding of alcoholism is as complex as the disease. In an interdisciplinary report, NOVA looks at the many faces of alcoholism—medical, historical and social.

Climate Crisis (The) 1019
December 20, 1983
Producer(s): Melanie Wallace, NOVA; Richard Broad, Thames Television
This summer's record temperatures may be one of the signs that the earth's atmosphere is warming up. NOVA looks at the climate predictions and hazard warnings for the next century, based on the effects of our soaring consumption of fossil fuels.

Miracle of Life (The) 1004
December 15, 1983
Producer(s): Bebe Nixon, NOVA ; Bo G. Erikson & Carl Lofman, SVT/Sweden
NOVA presents the first film ever made of the incredible chain of events which turns a sperm and an egg into a newborn baby. Amazing photographic techniques give the viewers the feeling of being reduced to the size of cells, following the sperm on its perilous voyage toward the egg, and meeting protectors and enemies along the way—like Ulysses on a microscopic odyssey.

Nuclear Strategy—A Guide for the Beginner 1018
December 13, 1983
Producer(s): Robert Zalisk, NOVA
Will nuclear weapons deter World War III or only make it more likely? NOVA explores the military strategies of the nuclear age, now that the challenge may no longer be to win global war but to prevent it.

Twenty-Five Years in Space 1017
December 6, 1983
Producer(s): Bebe Nixon, NOVA; Max Whitby, BBC
As the American space program celebrates its 25th anniversary this year, NOVA chronicles the effects of the space age on earth, drawing on popular music, film and television archives from the last quarter of a century.

Captives of Care 1016
November 29, 1983
Producer(s): Bebe Nixon, NOVA
Patients at an Australian institution for the severely handicapped rebel against a part of over-zealous custodians. This astonishing true story was filmed as a docudrama, written and performed by the patients themselves.

Normal Face: The Wonders of Plastic Surgery 1015
November 22, 1983
Producer(s): Theodore Bogosian, NOVA
When plastic surgeons repair the shattered face of a soldier or rescue a child from a disfiguring disease, the victory is more than skin-deep. NOVA looks at the history, heroes and miracles of plastic surgery in mending the accidents of war and birth.

To Live Until You Die: The Work of Elizabeth Kubler-Ross 1013
November 8, 1983
Producer(s): Eric Davidson, Andrew Maclear & Melanie Wallace, NOVA
Dr. Elizabeth Kubler-Ross has become a legend in her lifetime for her work with the dying. For the first time on American television, her explorations with patients are captured on film, as NOVA presents an intimate portrait of the Swiss-born psychiatrist at work.

Papua New Guinea: Anthropology on Trial 1012
November 1, 1983
Producer(s): Barbara Gullahorn-Holecek, NOVA
Remote tribes and exotic islanders have been made known to the world through the lens of anthropology. But in recent years, some of these people have begun to object. NOVA travels to Margaret Mead's Papua New Guinea and looks at anthropology from the other side.

Talking Turtle 1011
October 25, 1983
Producer(s): Melanie Wallace, NOVA; Suzanne Campbell Jones, BBC
NOVA looks at computers in the classroom through the eyes of MIT's Seymour Papert, father of the Turtle—a computerized robot that crawls on the floor and talks in a versatile language even five-year-olds can learn.

Artificial Heart (The) 1010
October 18, 1983
Producer(s): Stuart Harris, NOVA
Seattle dentist Barney Clark received the first complete artificial heart implant in 1982 and lived on for three post-operative months. NOVA investigates the risk, costs and controversies surrounding the development of the artificial heart.

Signs of the Apes, Songs of the Whales 1009
October 11, 1983
Producer(s): Linda Harrar, NOVA
The dream of talking with animals has been with us for centuries. NOVA explores the latest research, from language experiments with dolphins and apes to studies of animal calls in the wild. [11th season premiere]

Sixty Minutes to Meltdown 1008
March 29, 1983
Producer(s): Brian Kaufman, NOVA
The accident at Three Mile Island made front page news all over the world and rocked the entire nuclear power industry. In this special 90-minute broadcast, NOVA presents a docudrama chronicling the minute-by-minute events leading up to the accident and examines the questions raised about safety issues confronting the nuclear power industry today.

Fat Chance in a Thin World 1007
March 22, 1983
Producer(s): Thea Chalow, NOVA
"Why can't I lose weight?," It's a question many Americans ask themselves everyday. NOVA comes up with some surprising answers about weight and dieting that could have significant impact on our daily lives.

City of Coral 1006

March 8, 1983

Producer(s): Neil Goodwin, NOVA

NOVA takes a spellbinding voyage through one of the world's most fascinating and colorful ecosystems: a coral reef, where the line between plants and animals is blurred, "rocks" move, eat and fight, fish farm, and weak animals borrow the shields and weapons of stronger ones.

Asbestos: A Lethal Legacy 1005

March 1, 1983

Producer(s): Melanie Wallace, NOVA; John Willis, Yorkshire Television

Every 58 minutes between now and the end of the century, one American will die from asbestos exposure. NOVA turns its spotlight on the tragic consequences of asbestos use and on the current controversy over who is responsible.

Lassa Fever 1003

February 8, 1983

Producer(s): Alain Jehlen, NOVA ; Ruth Caleb, BBC

A gripping docudrama about a mysterious, highly lethal disease which struck a village in Nigeria in 1969, and the frustrating, seesaw battle against it. NOVA recounts how public health workers came perilously close to accidentally releasing a deadly virus in the US.

Pleasure Of Finding Things Out (The) 1002

January 25, 1983

Producer(s): Bebe Nixon, NOVA ; Christopher Sykes, BBC

NOVA captivates a remarkably candid portrait of Nobel prize-winning physicist Richard Feynman, a man of few pretensions and tremendous personal charm, who speaks with the same passion about a child's toy wagon and the frontiers of subatomic physics.

Hawaii: Crucible of Life 1001
January 18, 1983
Producer(s): Bebe Nixon, NOVA; Roger Jones, BBC
This land of fire and beauty is the most isolated island chain in the world. NOVA cameras uncover an extraordinary world far from the teeming tourist hotels, one filled with unique life forms, but also scarred by tragic extinction.

Making of a Natural History Film (The) 919
January 4, 1983
Producer(s): Mick Rhodes, BBC
To celebrate its 10th broadcast season, NOVA repeats the very first NOVA program every aired, a fascinating and delightful program about how wildlife films are made.

Tracking The Supertrains 918
December 14, 1982
Producer(s): Theodore Bogosian, NOVA
While America's passenger-train service deteriorates, trains in Japan and Europe are speeding ahead at over 150 miles per hour. NOVA reports that the super-fast trains are finally coming to America.

Whale Watch 917
December 7, 1982
Producer(s): Melanie Wallace, NOVA; Franz Lazi Films
NOVA follows the great grey whales along their annual marathon migration from the Arctic to the Mexican coast and reveals little known facts about the mating and feeding habits of the gentle giants.

Cobalt Blues (The) 915
November 23, 1982
Producer(s): Theodore Bogosian, NOVA
An investigative report on US dependence on foreign sources of strategic minerals, vital to the aerospace and steel industries, which

examines and questions the Reagan Administration policies toward those international sources.

Adventures of Teenage Scientists 914
November 16, 1982
Producer(s): Linda Harrar, NOVA
NOVA introduces some of the winners of the 1982 Westinghouse Science Talent Search: high school students whose interests range from silkworms to solar cells. With education facing a deepening financial crisis, will this year's group of well-trained young scientists be among the last of the best and the brightest?

Here's Looking At You Kid 913
November 9, 1982
Producer(s): Theodore Bogosian, NOVA; Andrew McGuire, The Burn Council
Of the 70,000 Americans hospitalized annually for severe burns, one-third are children. NOVA tells the story of extraordinary personal resilience in an 11-year-old boy's fight to recover from burns suffered over 73 percent of his body.

Goodbye Louisiana 916
November 3, 1982
Producer(s): Barbara Gullahorn-Holecek, NOVA
NOVA reports on the staggering water problems of Southern Louisiana—where the mighty Mississippi is threatening to change its course, and where last year 49 square miles of coastline disappeared into the Gulf of Mexico.

Fragile Mountain (The) 912
October 19, 1982
Producer(s): Sandra Nichols, NOVA
The Himalayas, highest peaks in the world, are crumbling. People are making them crumble, and people are the victims, as NOVA reveals in this breathtaking documentary.

Case Of the UFOs (The) 911
October 12, 1982
Producer(s): John Groom, BBC
For the first time on television a rigorous, scientific investigation into the fact, fiction, and hoax of unidentified flying objects. With vivid film and accounts from several eyewitnesses including astronauts, NOVA sifts the evidence for and against the existence of UFOs. [10th season premiere]

Aging: The Methuselah Syndrome 910
March 28, 1982
Producer(s): Sari Sapir, NOVA
What is aging? Why does it happen? Can it be stopped? NOVA presents a startling report on research into the processes which make us age and how to control them.

Animal Imposters 909
March 14, 1982
Producer(s): John Borden & Neil Goodwin, NOVA
In this vivid study of mimicry and camouflage NOVA takes a dramatic look at how snakes, butterflies, fish, turtles and many other kinds of animals, both predators and their intended victims, use remarkable forms of deception to achieve their goal: to eat, or avoid being eaten.

Palace of Delights 908
March 7, 1982
Producer(s): Bill Couturie & Jon Else, NOVA
NOVA visits San Francisco's Exploratorium—part laboratory, part school, part three-ring circus—run by an unlikely collection of physicists and high school students.

Life: Patent Pending 907
February 28, 1982
Producer(s): Robert Zalisk, NOVA
NOVA shows how scientists go about creating new forms of life,

and investigates the impact of the gene bonanza on industry, medicine, and the universities themselves. NOVA reveals that other countries are plowing far more resources than the US into the burgeoning industry.

Television Explosion (The) 906
February 14, 1982
Producer(s): Thea Chalow, NOVA
NOVA explores the past, present, and future of American television including the potential of cable, the Columbus, Ohio two-way TV experiment, the array of new techniques and their potential social impact. Will the new video technology let people see what they really want, rather than what the networks want?

Finding A Voice 905
February 7, 1982
Producer(s): Martin Freeth, BBC
What is it like not to be able to communicate with others? NOVA explores the severest of speech disabilities with Dick Boydell—born with cerebral palsy, confined to a wheel chair and unable for 30 years to say more than "yes" or "no" and investigates some of the new technology that gives the speechless a "voice."

Hunt For The Legion Killer (The) 904
January 31, 1982
Producer(s): Linda Harrar, NOVA; Dominic Flessati, BBC
One of the biggest investigations in medical history began when a mysterious killer disease broke out during independence celebrations in Philadelphia in 1976: Legionnaire's Disease. NOVA traces the search for a cause and cure—a search bedeviled by false trails, accusations of incompetence and cover-up, and increasing urgency as the death toll mounted.

Field Guide to Roger Tory Peterson (A) 903
January 24, 1982
Producer(s): James Murray, NOVA
NOVA takes an intimate look at Robert Tory Peterson, the man whose best-selling guide books to ornithology have played a pivotal role in turning birdwatching into a mass sport.

Test-Tube Babies 902
January 17, 1982
Producer(s): Peter Williams, NOVA & TVS
NOVA presents a dramatic, exclusive film of the first "test-tube" baby born in America, Elizabeth Jordan Carr. NOVA follows the pregnancy from the start, presenting the only view on American TV of the extraordinary medical procedures used to remove and fertilize the egg, and of the historic birth, December 28, 1981 in Norfolk, VA.

Salmon on the Run 901
January 10, 1982
Producer(s): Lynn Adler, Steve Christiansen & Jim Mayer, NOVA
NOVA captures the breathtaking power and determination of these amazing creatures and examines how business and technology are changing the fishing industry—and the salmon itself.

Twins 820
December 6, 1981
Producer(s): Linda Harrar, NOVA ; Heather Cook, CBC
Ever thought about what it would be like to have your mirror image talk back to you? It can be an everyday occurrence for identical twins. NOVA tells the incredible story of scientific research on twins—a field marked by brazen and damaging fraud, but also by surprising and important new discoveries about nature's recipe of heredity and environment which makes us all unique individuals.

City Spaces, Human Places 819
November 29, 1981
Producer(s): Pat Kent, NOVA; William Whyte, MAC
William H. Whyte's insightful and humorous look at city parks, plazas and streets, and the people who use them, showing the remarkable research he did over a period of many years to find out why some city squares and small parks are enjoyable while others are so dreary. His work led to the transformation of some New York City plazas from barren to bustling. Whyte shows how any city—large or small—can lick the problem of downtown dreariness.

Notes of a Biology Watcher: A Film With Lewis Thomas 818
November 22, 1981
Producer(s): Robin Bates, NOVA
You are not alone! Like it or not, every human being and virtually every living creature is, in a sense, owned and operated by legions of prehistoric organisms, hordes of them in each cell in the body. That is one of the startling revelations as NOVA explores the mysterious wonder of life with Dr. Lewis Thomas, a leading biologist and award-winning author described by Time as "quite possibly the best essayist on science anywhere in the world."

Artists in the Lab 817
November 15, 1981
Producer(s): Barbara Gullahorn-Holecek, NOVA
Many were delighted by the extraordinary special effects in movies like "2001" and" Star Wars," but few realized how their magic relied on technologies as futuristic as their science fiction plots. NOVA introduces 20th century pioneers who use computers and lasers to create an extraordinary array of strange, exciting new art forms.

Did Darwin Get It Wrong? 816
November 1, 1981
Producer(s): Theodore Bogosian, NOVA ; Alec Nisbett, BBC
The controversy which exploded a century ago when Charles Dar-

win published "The Origin of Species" is erupting again with new facts and emotion. NOVA explores challenges to the theory of evolution coming from evidence in fossils, from biology laboratories, and Creationists.

Locusts: War Without End 815
October 25, 1981
Producer(s): Mick Rhodes, NOVA
Called the "teeth of the wind" by those who have battled them for centuries, locusts continue to plague hundreds of millions of people and their land. Rare desert rains transform locusts from harmless grasshoppers to voracious swarms capable of destroying all vegetation in their path. NOVA reveals some of man's latest attempts to rid himself of his age-old enemy, the locust.

Cosmic Fire 814
October 18, 1981
Producer(s): Robert Zalisk, NOVA
NOVA studies the extraordinary discoveries of X-ray astronomy. This new science has revealed that our universe is much stranger and more violent than ever imagined, filled with neutrons, stars, exploding galaxies, quasars and black holes—a universe seething with energy, bursting across vast distances of space and time.

Great Violin Mystery (The) 813
October 11, 1981
Producer(s): Bebe Nixon, NOVA; Barry Stoner, WHA-Madison
A great secret lies locked inside the master violins created by Italian craftsmen like Antonio Stradivari in the 17th and 18th centuries. Now, a Wisconsin physicist, working alone in his cellar, may have solved the violin mystery.

Why America Burns 812
October 4, 1981
Producer(s): Brian Kaufman, NOVA
More people die in fires in the US than in any other industrialized country. In an alarming report that challenges the complacency of the US fire prevention establishment, NOVA uncovers glaring gaps in our defenses against flames that kill. Sealing any one of these gaps might save thousands of lives and prevent enormous pain and misery.

Computers, Spies & Private Lives 811
September 27, 1981
Producer(s): Theodore Bogosian & Linda Harrar, NOVA
NOVA reports on the potential danger of modern computers that gather "routine" information about our daily lives as we buy things, go to the hospital, or make donations. Computers can know more about us than our closest friends. NOVA examines how much of that personal information is readily shared with other computers. [9th season premiere]

Resolution on Saturn 810
August 28, 1981
Producer(s): Fisher Dilke, BBC
It's over 300 years since Galileo turned his new telescope on Saturn and first saw its spectacular rings. NOVA shows the beauty and new mysteries discovered by Voyager 1 on its historic visit.

Animal Olympians 809
March 17, 1981
Producer(s): Jeffrey Boswall, BBC
The beauty, endurance, and raw power of animals in the wild are captured on film as NOVA juxtaposes Olympic athletes performing feats which have parallels in the animal kingdom with animals who are the champions of grace and strength.

Asteroid and the Dinosaur (The) 808
March 10, 1981
Producer(s): Robin Bates, NOVA
For 150 million years, dinosaurs dominated the earth. Then, 65 million years ago, they suddenly vanished, along with a great deal of the planet's animal and plant life. NOVA examines a remarkable theory about the cause of the catastrophe—in which the first clue to the solution was a piece of clay.

Beyond The Milky Way 807
March 3, 1981
Producer(s): Alec Nisbett, BBC
Sophisticated instruments used by astronomers enable earthlings to see beyond what was once the cloudy barrier of the Milky Way, to a universe of perhaps 100 billion other galaxies. NOVA takes a trip into outer space to see these clusters which are as old as time and several million light years away.

Malady Of Health Care (The) 806
February 24, 1981
Producer(s): Barbara Gullahorn-Holecek, NOVA
Health care is no longer two aspirins and some chicken soup—it is a huge enterprise capable of amazing feats and costing billions of dollars. How can we afford to pay the bills? Is quality health care a right or a privilege? NOVA examines these questions in a comparison between the American and British systems of health care.

Science Of Murder (The) 805
February 17, 1981
Producer(s): Brian Kaufman, NOVA
NOVA investigates what science can do in helping to solve murder—in understanding why it occurs, and how the rate might be reduced—and explores the work of people who have the stark job of dealing with death: the police, pathologists, scientists and psychiatrists.

Anatomy of a Volcano 804
February 10, 1981
Producer(s): Brian Kaufman, NOVA; Stuart Harris, BBC
When Mt. St. Helens erupted earlier this year, it focused the attention of the whole world on the almost incredible destructive forces that volcanos can release. Geologists from around the world congregated at the volcano and NOVA joined the vigil for an in-depth look at the incident and its aftermath.

The Dead Sea Lives 803
January 27, 1981
Producer(s): Pat Kent, NOVA; Christopher Riley, BBC
NOVA examines the Dead Sea. The lowest place on earth, at 1400 feet below sea level, it is jointly owned by Israel and Jordan. If used properly it could become a vital natural resource for both countries, giving them not only salt, but protein, fertilizer, oil, and a solar energy store.

Message In The Rocks 802
January 20, 1981
Producer(s): Linda Harrar, NOVA; Alec Nisbett, BBC
This program explores clues gathered from ancient rocks and meteorites in an attempt to piece together how our planet formed, what happened during its earliest days, and when life first appeared. The program includes visits to the scene of a fresh fall of meteorites, several volcanic eruptions, and an underwater glimpse of molten "pillow" lava as it oozes out of volcanic vents in the sea floor.

Doctors Of Nigeria (The) 801
January 6, 1981
Producer(s): Barbara Gullahorn-Holecek, NOVA
Is the fagara root a match for the stethoscope? This program looks at the contributions of both traditional herbal medicine and western orthodox medicine to the health of the Nigerian people.

It's About Time 720
December 30, 1980
Producer(s): Chris Haws, BBC
Time—a concept which has baffled scientists and philosophers since time immemorial. Actor Dudley Moore hosts a funny, sobering and visually stunning quest for answers to riddles, as NOVA spends an hour on time.

Red Deer of Rhum (The) 719
December 23, 1980
Producer(s): Robert Zalisk, NOVA ; Peter Jones, BBC
The cuddly image of Rudolph the Red-Nosed Reindeer has become an integral part of the jollity of the Christmas season. NOVA takes a timely look at how real deer live by visiting Rhum—an island off the coast of Scotland inhabited by red deer.

Touch of Sensitivity (A) 718
December 9, 1980
Producer(s): Pat Kent, NOVA; Stephen Rose, BBC
The exquisite sensitivity of tough cells in the human skin makes it possible for us to discriminate with precision the slightest changes in texture and pressure, but how the electrical impulses we receive are converted into sensation remains a mystery. NOVA explores the hidden meaning and extraordinary power of human touch.

Moving Still 717
December 2, 1980
Producer(s): Tony Priano, NOVA; Chris Hawes, BBC
NOVA tells the story of still and cine photography in science—from the extraordinary work of the pioneers in the early 1800s to how the ability to freeze time on film in ever shorter periods has given scientists remarkable new insights. Today photography enables us to analyze (frame by frame) the thousands of molecular reactions that can happen in less time than the blink of an eye.

Water Crisis (The) 716
November 25, 1980
Producer(s): Sari Sapir, NOVA
NOVA travels to the Adirondack Mountains where acid rain is killing many high elevation lakes; to the Mississippi River where chlorine has combined with natural and manmade organic chemicals to form cancer-causing toxic chemical substances; to California, where conservation recycling has had to become a way of life; and to Bedford, Massachusetts, where the town wells have been contaminated by industrial waste.

Wizard Who Spat On The Floor (The) 715
November 18, 1980
Producer(s): Theodore Bogosian, NOVA; Robert Vas, BBC
Thomas Edison is the quintessential American hero, the Wizard whose inventions revolutionized modern living. But there was always more to Edison than met the eye. He was a complex and contradictory man; a brilliant inventor, a foolish investor; a demanding boss, a liberal benefactor—a public figure that no one ever really knew. NOVA profiles the man behind the mythical reputation.

Voyager: Jupiter & Beyond 714
November 11, 1980
Producer(s): Robert Zalisk, NOVA; Fisher Dilke, BBC
On Wednesday, November 12, 1980, Voyager 1 is expected to arrive at Saturn for a first time ever extensive close-up investigation of the majestic ringed planet. Astronomers can expect to gather more information than ever before possible. On the day before this historic event, NOVA documents Voyager's journey through the outer solar system.

Big IF (The) 713
November 4, 1980
Producer(s): Theodore Bogosian, NOVA; Vivienne King, BBC
Is interferon—known as IF in medical shorthand—the wonder drug and cure for cancer that some doctors claim? NOVA travels to Lon-

don, Stockholm, Houston, San Francisco, and New Haven in search of the answer in the most complete film on interferon ever to appear on American television.

Do We Really Need The Rockies? 712
October 28, 1980
Producer(s): Robin Bates, NOVA
Locked in the shale of the Western Rocky Mountains is more oil than in the Middle East—more than enough to solve our dependence on foreign crude oil. But will shale oil solve our gasoline shortage, or will it simply turn the Rockies into a gigantic industrial zone? NOVA explores the promise and the problems of shale oil.

Sea Behind The Dunes (The) 711
October 14, 1980
Producer(s): John Borden & Neil Goodwin, NOVA
One year in the intricate life of a coastal lagoon unfolds in an hour's time when NOVA documents the fragile tidal ecosystem which supports the entire ocean.

Cancer Detectives Of Lin Xian (The) 710
October 7, 1980
Producer(s): Robert Zalisk, NOVA; Edward Goldwyn, BBC
In one of the first films ever to come out of modern China, NOVA sifts through clues that Chinese scientists have uncovered in their pursuit of particularly virulent and elusive forms of cancer from which one out of every four people die.

The Pinks and the Blues 709
September 30, 1980
Producer(s): Veronica Young, NOVA
NOVA explores the shaping and molding of the male and female personality. From infancy through childhood, the program documents the impact of culture on the development of sex differences. [8th season premiere]

Mr. Ludwig's Tropical Dreamland 708
March 25, 1980
Producer(s): Joan Freeman, NOVA; Vivienne King, BBC
NOVA explores the amazing Jari project of the Amazon basin. Eleven years ago, 3.5 million acres of virgin jungle were bought by the reclusive billionaire, Daniel K. Ludwig.

Mediterranean Prospect (A) 707
March 18, 1980
Producer(s): Joan Freeman, NOVA; Peter Jones, BBC
Every year, millions of tourists converge on the Mediterranean's sunny coasts, lured by the prospect of bathing in clear, azure waters and basking in semi-tropical sun. But years of use and abuse have taken their toll on the once idyllic Mediterranean and the "world's biggest swimming pool" has become the world's biggest open sewer. NOVA explores the complex problems that plague the Mediterranean's future.

Safety Factor (The) 706
March 11, 1980
Producer(s): Paula Apsell & Alec Nisbett, NOVA & BBC
Recent aircraft accidents have raised the question of just how safe modern commercial aviation really is. NOVA looks at some of the problems and experimental efforts underway to deal with them.

Umealit: The Whale Hunters 705
March 4, 1980
Producer(s): John Angier, NOVA
Whaling is an integral part of Eskimo life, and a major source of food; even so, conservationists are seeking to restrict the hunting of bowheads in Alaska.

Portrait of a Killer 704
February 19, 1980
Producer(s): Bo G.Erikson, Carl Lofman & Sari Sapir, NOVA & SVT
More than 40 million Americans are afflicted by cardiovascular disease. NOVA examines the new information on risk factors and possible prevention of heart attacks and strokes—often fatal diseases.

Living Machines 703
February 5, 1980
Producer(s): Robin Bates, NOVA
NOVA explores the science of natural engineering and asks the basic questions: what makes a good design in nature and why did a particular plant or animal adopt a particular design?

A is for Atom, B is for Bomb 702
January 22, 1980
Producer(s): Brian Kaufman, NOVA
NOVA profiles Dr. Edward Teller, the "Father of the Hydrogen Bomb," an acclaimed scientific genius and brilliant theoretician, and a man considered by some the most dangerous scientist in the United States.

Elusive Illness (The) 701
January 15, 1980
Producer(s): Robin Bates, NOVA
Aborigines in Australia, woodchucks in Pennsylvania, the Nobel Prize in Stockholm and the gay community in New York—what could possibly link such disparate elements? The answer is Hepatitis. NOVA examines this elusive disease, what causes it, how it is spread and how you get rid of it. [7th season premiere]

Blindness: Five Points of Views 619

December 18, 1979

Producer(s): Veronica Young, NOVA

For many people the idea of life without vision is as fearful as death. NOVA looks at five people struggling to save their threatened vision using drugs, surgery, counseling and determination.

Termites and Telescopes 618

December 11, 1979

Producer(s): Brian Kaufman, NOVA ; Peter Jones, BBC

Dr. Philip Morrison, Institute Professor and Professor of Physics at Massachusetts Institute of Technology, presents this thoughtful and provocative commentary on the nature of civilization.

The Bridge That Spanned the World 617

December 4, 1979

Producer(s): Brian Kaufman, NOVA; Bob Bootle, BBC

The Iron Bridge across the River Severn in Telford, England is two centuries old this year. It remains a monument to the Shropshire iron masters who built it, and a symbol of the Industrial Revolution that was born in the area where the bridge stands. NOVA traces the development of ironmaking and its far-reaching effects on society and the world economy.

India: Machinery of Hope 616

November 20, 1979

Producer(s): Barbara Gullahorn-Holecek, NOVA

Most of India lives by the same rhythm and uses the same tools, as in centuries past. But there is another India—with thriving commercial centers, spotless research laboratories and large-scale industries. NOVA looks at how the gap between these two extremes is shrinking because of a policy of "appropriate" technology that uses the resources of both to meet the greatest needs of all.

All Part of The Game 615
November 6, 1979
Producer(s): Paula Apsell, NOVA
Thousands of amateur athletes are hurt every year, and many professional athletes suffer injuries that may mean the end of a career. NOVA looks at a new medical specialty—sports medicine—which promises to prevent and cure many sports related problems.

Race For The Gold 614
October 30, 1979
Producer(s): Paula Apsell, NOVA
At the 1976 Olympics, East German athletes walked off with 40 of the coveted gold medals, though their country is only the size of New Jersey. NOVA investigates whether a drug is responsible for their incredible success—or is American athletic training and commitment falling behind that of the Communist world?

Sweet Solutions 613
October 16, 1979
Producer(s): Theodore Bogosian, NOVA; Vivienne King, BBC
NOVA views the history of sugar—from its scientific, religious and political history to its medical controversy.

Life On A Silken Thread 612
October 9, 1979
Producer(s): Sari Sapir, NOVA; Jorg Dattler, German TV
Sinister, sometimes even deadly, spiders have little popular appeal; yet their silken webs are among nature's loveliest creations. NOVA takes a close-up look in slow motion, as spiders reveal a delicate grace and beauty, and an amazing array of lifestyles.

Plague on our Children (A) 611
October 2, 1979
Producer(s): Robert Richter, NOVA
Is the chemical industry a boom to modern civilization, or a major

threat to our health and that of future generations? NOVA examines how toxic herbicides, pesticides, and other chemicals may cause cancer, miscarriages and birth defects in humans.

Keys Of Paradise (The) 610
March 29, 1979
Producer(s): Dick Gilling, BBC
Some powerful painkilling drugs have just been discovered—in a place where you would least expect to find them. Endorphins and their component enkephalins are manufactured in the brain, and perform the same painkilling function as analgesics like morphine. NOVA explores some physiological mysteries, such as why acupuncture works, how placebos can relieve symptoms, and how endorphins could revolutionize the treatment of pain, depression, and even schizophrenia.

Einstein 609
March 15, 1979
Producer(s): Patrick Griffin, NOVA
One hundred years after his birth, Albert Einstein remains an enigma to most Americans. NOVA presents an insightful portrait of the man and his mind through rarely viewed film footage.

Beersheva Experiment (The) 608
March 8, 1979
Producer(s): Sari Sapir, NOVA; Edward Goldwyn, BBC
Health care is the third largest industry in the US. As a result of billions of dollars spent for medical education in the 1960s, there are now too many specialists and too few primary care physicians, especially in undeserved areas. NOVA tells the story of one medical school in Israel that is training a new kind of family doctor.

End Of The Rainbow (The) 607
March 1, 1979
Producer(s): Brian Kaufman, NOVA
Is nuclear fusion the solution to the energy crisis? NOVA examines the promise—and problems—of fusion as a future energy source.

Invisible Flame (The) 606
February 22, 1979
Producer(s): Alec Nisbett, BBC
Some day hydrogen may replace the gasoline that we are now using up so rapidly. NOVA looks at the potential of hydrogen as a zero-pollution fuel.

Patterns From The Past 605
February 8, 1979
Producer(s): Sari Sapir, NOVA
Below the snow-capped peaks of the Peruvian Andes, the Q'eros Indians live a life patterned on that of their ancestors thousands of years ago. NOVA takes a look at the unchanging world of these isolated mountain people.

Cashing In On The Ocean 604
February 1, 1979
Producer(s): Stuart Harris, NOVA
The bed of the northeast Pacific Ocean was covered with a "carpet" estimated to be worth a staggering ten million dollars. These manganese nodules—the bumpy carpet—are rich not only in manganese but in the key strategic minerals: copper, nickel and cobalt. NOVA examines the debate about who owns them and has the right to exploit their use.

World of Difference (A) 603
January 18, 1979
Producer(s): Veronica Young, NOVA
In 1945, B.F. Skinner shocked the world by putting his 13 month-old daughter, Deborah, into a "box." The box was actually a climate-con-

trolled crib designed for comfort and protection, and the young psychologist was merely testing his theory that environment controls behavior. NOVA portrays the life of this famous behavioral psychologist now in his 70's and living quietly in Cambridge as Emeritus Professor of Psychology at Harvard University.

Long Walk Of Fred Young 602
January 11, 1979
Producer(s): Michael Barnes, BBC
As a child, Fred Young hunted birds and wild animals with primitive weapons, spoke only the Indian languages Ute and Navajo, went to a medicine man when he was sick, and slept under the stars. NOVA profiles Dr. Frederick Young, now a nuclear physicist working on the laser fusion project at the Los Alamos Scientific Laboratory in New Mexico.

Black Tide 601
January 4, 1979
Producer(s): Graham Chedd, NOVA
On March 16, 1978, the US owned, Liberian registered supertanker Amoco Cadiz went aground off the coast of Brittany. Over the ensuing weeks its entire 68 million gallons of oil drained into the sea. A NOVA team began filming at the scene shortly after the disaster, the biggest oil spill in history, to record clean-up efforts, effects of the spill on the crucial tourism and fishing industries, and the attempts of marine biologists to trace the passage of the oil through the environment.

Alaska: The Closing Frontier 520
June 28, 1978
Producer(s): Paula Apsell, NOVA
Congress is currently considering a proposal that would double the size of America's national park system by designating a sizeable chunk of Alaska as off-limits to developers. NOVA explores the public debates on Alaska before it, such as the construction of the oil pipeline—a proposal that has sparked a bitter controversy between conservationists and developers.

Whisper From Space (A) 519
June 21, 1978
Producer(s): Peter Jones, BBC
In 1965, Arno Penzias and Robert Wilson, two radio astronomers at Bell Telephone Laboratories, discovered faint, but ever-present, microwave signals from space—the most ancient and most distant signals detected by man: the oldest "fossils" in the universe. NOVA explores the current surge of cosmological discovery that continues to aid scientists in the "cosmic archaeology" of digging into the history of the universe.

Memories From Eden 518
June 14, 1978
Producer(s): Barbara Gullahorn-Holecek, NOVA
Traditional zoos were designed neither for people nor animals; cages taught people more about their separation from nature than about an animal and its habitat. But just as man has realized that he has all but destroyed much of the world's wilderness and its wildlife, he is realizing that the zoo may be a last refuge. NOVA visits U.S. zoos to examine issues of concern today: breeding, public education, creative animal habitats, and reintroducing animals to their native habitats.

Tse Tse Trap (The) 517
June 7, 1978
Producer(s): Edward Goldwyn, BBC
NOVA explores Bovine sleeping sickness. Spread by a fly, it is a deadly disease that poses a threat to Africa's cattle.

Desert's Edge (The) 516
May 31, 1978
Producer(s): Sari Sapir, NOVA; Richard Taylor, BBC
For thousands of years people have managed to live in deserts all over the world. But in recent years, a growing population and the demands of the international market have put more stress on these poor and

easily exhausted lands. NOVA examines the consequences and possible solutions to desertification.

Insect Alternative (The) 515
May 24, 1978
Producer(s): Graham Chedd, NOVA
Some form of pest control is desperately needed in a world that loses up to 40 percent of its crops to insects each year. Chemical pesticides have backfired. Pesticide-resistant insects frequently develop, and previously harmless insects have become devastating infestations. Farmers have found themselves trapped on a "pesticide treadmill"—the more they spray, the more they need to spray. NOVA examines several alternatives for pest control.

Light of the 21st Century 514
May 10, 1978
Producer(s): Tony Edwards, BBC
When first invented 18 years ago, lasers were called "a solution looking for a problem; " nobody could think what to do with them. But in fact research scientists immediately began to exploit their pure colors and near-perfect focusing ability. Today lasers are a billion-dollar business, used in construction, manufacturing, dentistry and medicine. And the future uses of lasers are likely to be of major significance in nuclear fusion and communications.

Road To Happiness (The) 513
May 3, 1978
Producer(s): Patrick Griffin & Francis Gladstone, NOVA
Henry Ford, a great friend of Edison, was a film enthusiast who amassed some one and a half million feet of film during his lifetime. Deposited in the National Archives and known as the Ford Film Collection, it covers not only the Ford family and Ford Motor Company but also contains newsreels, and general films produced under Ford. Using the Collection, NOVA profiles Ford's life and times.

Battle for the Acropolis 512
April 19, 1978
Producer(s): Roy Davies, BBC
The fortified plateau above Athens known as the Acropolis, including the Parthenon, is the site of some of the most remarkable architecture in the world, representing the artistic peak of classical Greek architecture. NOVA examines how the heavily polluted air of Athens produces acid rain which is dissolving the marble sculptures and columns; and how iron tiles used extensively in repair 40 years ago are now rusting, expanding and shattering the stone structures.

Still Waters 511
April 12, 1978
Producer(s): John Borden & Neil Goodwin, NOVA
NOVA shows a year in the life of a beaver pond and includes almost every life form that exists in, on, under, around and above the water, from the microscopic plant life of summer to the eagles feeding on carcasses of deer that collapsed on the winter ice.

Icarus' Children 510
March 29, 1978
Producer(s): Simon Campbell Jones, BBC
In the summer of 1977 Paul MacCready, a California scientist and businessman, won the coveted Kremer Prize. His achievement was to design and build an airplane which completed, unaided, a one-mile figure-eight course entirely under the power provided by the pilot himself. This is the story of those many failures and MacCready's success.

Mind Machines (The) 509
March 22, 1978
Producer(s): Paula Apsell, NOVA
Today's scientists may be creating their own successors. Work being done in Artificial Intelligence (AI), a branch of computer science,

only suggest that in the not too distant future, machines will outpace their creators. NOVA examines the possibility.

Case Of The Ancient Astronauts, The 508
March 8, 1978
Producer(s): Graham Massey, BBC
NOVA investigates the theories of von Daniken and others that the Earth has been visited by intelligent beings from outer space. Among claims examined are: that the building techniques used in the Great Pyramid of Cheops are so advanced that only an extraterrestrial intelligence could have built it; and that the engraved stones of Palenque in Mexico depict an ancient astronaut at the controls of a space rocket.

Great Wine Revolution (The) 507
March 1, 1978
Producer(s): Dominic Flessati, BBC
A science-based revolution in the making of wine is underway. NOVA traces the secrets of the aging process and science's involvement with the prediction of mass produced, high-quality vintage wines.

Trial of Denton Cooley 506
February 22, 1978
Producer(s): Francis Gladstone, NOVA
In a dramatic docudrama, NOVA reconstructs the controversial lawsuit raised against renowned heart surgeon Dr. Denton Cooley when one of his patients died after heart surgery, and examines the legal and moral issues this raises in the practice of modern medicine.

Bamiki Bandula: Children of the Forest 505
February 15, 1978
Producer(s): Kevin Duffy, NOVA
In the rain forests of Zaire, in the heart of Africa, live the Mbuti Pygmies. The Pygmy way of life has always been extraordinarily difficult

to capture on film, though many have tried. NOVA presents a rare portrait of an elusive people, made by an independent filmmaker who lived with the Pygmies and won their trust.

Final Frontier (The) 504
February 1, 1978
Producer(s): Graham Chedd, NOVA
Second of the two-part series on space programs, NOVA looks ahead to the future, post-Apollo and the role that man in space will play, including the possibility of space colonization—huge orbiting space stations where people live and work in an earth atmosphere under artificial gravity.

One Small Step 503
January 25, 1978
Producer(s): Graham Chedd, NOVA
Part one of a two-part series on the subject of man in space, NOVA examines the history of NASA—from the origin of the space race through the triumph of the Apollo programs. By tracing the history of three key programs—Mercury, Gemini, Apollo—we show how the basic challenges surrounding space flight were answered: rendezvous and docking, life support, weightlessness, space sickness, equipment reliability and more.

Blueprints in the Bloodstream 502
January 18, 1978
Producer(s): Vivienne King, BBC
It has been known since the turn of the century that there are four human blood groups, based on different red cells and serum characteristics. NOVA looks at the more recent discovery that the different white cell types, as determined by a variety of different molecular markers on the cell surface, open up the possibility of the prevention of disease.

Green Machine (The) 420
January 11, 1978
Producer(s): Tony Edwards, BBC
Botany is a neglected science and plants are all around us, but unfamiliar. NOVA examines our state of knowledge of how plants work: growth hormones, responses to light and shade, photosynthesis, root mechanisms and twining responses.

In The Event of Catastrophe 501
January 4, 1978
Producer(s): Barbara Gullahorn-Holecek, NOVA
Can a nuclear war be survived? Some members of the defense community say yes. NOVA explores the possibility. [5th season premiere]

New Healers (The) 419
June 29, 1977
Producer(s): Paula Apsell, NOVA
NOVA explores the debilitating diseases that are often caused by poverty and follows two paths to health care in Tanzania and the United States.

Across the Silence Barrier 418
June 22, 1977
Producer(s): Francis Gladstone, NOVA
NOVA explores the different means by which hearing-impaired people have learned to penetrate the world of the hearing by visiting with Kitty O'Neil—a record-holding speed car racer; Frances Parsons, an advocate of hearing-impaired persons' rights; and workers at Silent Industries—a factory in Los Angeles founded by a deaf man.

Linus Pauling: Crusading Science 417
June 1, 1977
Producer(s): Robert Richter, NOVA
NOVA profiles Linus Pauling—the only person to have received two unshared Nobel Prizes for his work in nuclear weapons.

Tongues of Man, Part 2 416
May 18, 1977
Producer(s): Christopher Martin, BBC
In part two of its two-part series on the diversity of language, NOVA explores how man has coped with the confusion of language and asks if the growing acceptance of English is the answer.

Tongues of Men, Part 1 415
May 11, 1977
Producer(s): Christopher Martin, BBC
In part one of this two-part exploration of the diversity of world languages, NOVA examines how and why the bewildering confusion of languages came about.

Red Planet (The) 414
April 27, 1977
Producer(s): Tony Edwards, BBC
NOVA traces 300 years of speculation, investigation and discovery that have centered on Mars—particularly the theory that the planet could support life. Questions raised by NASA's 1976 Viking mariner missions about how the vast canyons were formed are also explored.

Dawn of the Solar Age (The) 412
April 20, 1977
Producer(s): Edward Goldwyn, BBC
Solar energy is increasingly popular as a home heating source. But only recently has it been seriously considered as a source of industrial power. NOVA looks at this new industrial approach, such as the use of a huge windmill in Ohio, giant machines that may generate electricity from the heat of the tropical seas or from the motion of waves, and an orbiting solar power station able to beam microwaves to earth.

Business of Extinction (The) 413
April 20, 1977
Producer(s): Barbara Gullahorn-Holecek, NOVA
NOVA explores the huge international illegal trade in animals, penetrates the thriving underworld of smugglers and assesses the effects on vanishing wildlife.

Wolf Equation (The) 411
March 30, 1977
Producer(s): Neil Goodwin, NOVA
In the winter of 1976-77, 80 percent of the wolf population in Northwest Alaska was the target of aerial hunts. Although the area is roamed by the Western Arctic caribou herds—a natural predator of the wolf—the caribou population has been steadily decreasing in number. NOVA examines how the Dept. of Fish and Game is handling the the problem of wolf control.

Human Animal (The) 410
March 23, 1977
Producer(s): Peter Jones, BBC
NOVA investigates the controversial theory of Harvard University biologist E.O. Wilson, that many aspects of human behavior are genetically determined.

Gene Engineer (The) 409
March 16, 1977
Producer(s): Graham Chedd & Paula Apsell, NOVA
NOVA explores the history of genetic engineering and the possible risks and benefits of this area of research.

Pill for the People (The) 408
March 9, 1977
Producer(s): Robin Brightwell, BBC
NOVA profiles chemist Russell Marker who made the birth control

pill possible by discovering a synthetic substitute for the hormone progesterone.

Bye Bye Blackbird 407
March 2, 1977
Producer(s): Peter Bale, NOVA
NOVA looks at blackbirds, their winter habit of nesting in the millions, and the destruction they do to crops.

Incident at Brown's Ferry 406
February 23, 1977
Producer(s): Robert Richter, NOVA
NOVA recreates March 1975 at Brown's Ferry, an Alabama nuclear power plant—the largest in the world—which suffered a seven-hour fire that came very close to developing into a major public disaster.

Plastic Prison (The) 405
February 9, 1977
Producer(s): David Kuhn, NOVA; Simon Welfare & Kevin Sim, Yorkshire Television
NOVA follows the lives of three boys who have Severe Combined Immune Deficiency—a disease that leaves its victims with no immune system.

Sunspot Mystery (The) 404
February 2, 1977
Producer(s): Ben Shedd & Graham Chedd, NOVA
NOVA explores the research on the 1976 drought in the western United States which led some solar scientists to discover the link between weather patterns and the 11 year sunspot mystery.

What Price Coal? 403
January 19, 1977
Producer(s): Francis Gladstone, NOVA
What is the price we are prepared to pay for coal? NOVA looks at the

environmental and health safety issues raised by the government, industry, and the victims.

Hot Blooded Dinosaurs (The) 402
January 12, 1977
Producer(s): Robin Bates & Robin Brightwell, BBC
If you were a dinosaur scientist, what would you do with a pile of fossil bones? How would you even start to put the giant jigsaw puzzle together, never mind discover anything about how these dinosaurs lived? NOVA explores the incredible world of the dinosaur scientist.

Hitler's Secret Weapon 401
January 5, 1977
Producer(s): Francis Gladstone & Patrick Griffin, NOVA
NOVA traces the development of Hitler's V-2 rocket through rare footage obtained from the National Archives—some never broadcast before on television. [4th season premiere]

Case of the Bermuda Triangle (The) 320
June 27, 1976
Producer(s): Graham Massey, BBC
Since 1945, hundreds of ships and planes and thousands of people have mysteriously disappeared in an area of the Atlantic Ocean off of Florida, known as the Bermuda Triangle. NOVA penetrates the mystery of the terrifying Bermuda Triangle.

Genetic Chance (The) 319
June 20, 1976
Producer(s): Robert Reid, BBC
Recent scientific developments have made it possible to detect a wide variety of defects in unborn babies. NOVA focuses on the ethical question that must be considered: What defines a defect? Should defective babies be aborted, or should they be allowed to live?

Inside the Shark 318
June 13, 1976
Producer(s): Tony Edwards, BBC
The "Jaws" phenomenon has given sharks a bad name. But is the shark really such a barbarian? NOVA looks at the lifestyle of this remarkable survivor from the days when dinosaurs ruled the earth.

Death of a Disease 317
June 6, 1976
Producer(s): Paula Apsell, NOVA
As late as 1967, smallpox struck as many as 15 million people in 43 countries and killed an estimated two or three million. Experts now believe that the disease is on the verge of extinction. NOVA looks at the recent success of the World Health Organization's program to eradicate this disease, considered a triumph of western-styled medicine.

Woman Rebel (The) 316
May 23, 1976
Producer(s): Francis Gladstone, NOVA
Margaret Sanger was responsible almost single-handedly for changing the whole attitude of the male-dominated medical profession towards "women's issues" and, above all, for gaining social and political acceptance for the concept of birth control. This NOVA docudrama reconstructs her life, told as flashbacks interspersed throughout an interview. Piper Laurie stars as Margaret Sanger.

Benjamin 315
May 9, 1976
Producer(s): Robin Bates & Graham Massey, BBC
Benjamin is a healthy, normal baby, whom we meet at birth and whose first year of life provides the backbone of this revealing NOVA study of early child development.

Hunters of the Seal 314
May 2, 1976
Producer(s): Barbara Gullahorn-Holecek, NOVA
NOVA revisits the once traditional Netsilik eskimos of Pelly Bay, ten years after the Canadian government imposed a Western lifestyle upon this ancient culture.

Underground Movement (The) 313
April 18, 1976
Producer(s): Suzanne Gibbs, NOVA
NOVA explores life underground, from foxes and badgers through moles and worms down to the myriad of micro-organisms that make soil the most complex substrate for life on earth. Included in the film is extraordinary footage of a mole burrowing and of roots growing.

Williamsburg File (The) 310
April 14, 1976
Producer(s): Antonia Benedek, BBC
NOVA joins chief archaeologist, Ivor Noel Hume, of Colonial Williamsburg, VA, for a fascinating glimpse at the lifestyles of the founders of this country, complete with detailed reconstructions of houses, stores, workshops, gardens, taverns and palaces.

Transplant Experiment (The) 312
April 11, 1976
Producer(s): Dominic Flessati, BBC
Dr. Norman Shumway of Stanford University has performed more heart transplants than any other heart surgeon. NOVA explores those extraordinary days in 1968–69 when it appeared that everyone with a scalpel was doing heart transplants, and the survival of patients was measured in days.

Overworked Miracle (The) 311
March 21, 1976
Producer(s): Christopher Riley, BBC
Today we take antibiotics for granted, and by so doing are steadily eroding their medical value. NOVA examines the problem of resistance to antibiotics in the bacteria they are designed to kill.

Race for the Double Helix (The) 308
March 7, 1976
Producer(s): Graham Chedd, NOVA
Author Isaac Asimov joins NOVA in the retelling of the remarkable story of the discovery of the structure of DNA. James Watson and his ex-colleague, Francis Crick, exchange memories of the events which led to their winning the race for the structure of the gene.

Renewable Tree (The) 309
March 7, 1976
Producer(s): Ben Shedd, NOVA
Each Sunday edition of the New York Times consumes 153 acres of trees. The paper packs, napkins, paper cups and packing used by McDonald's gobble up 315 square miles of trees every day. NOVA asks if, at this rate, trees can remain a renewable resource.

Ninety Degrees Below 307
February 15, 1976
Producer(s): David Kuhn & Franz Lazi, NOVA
There's one place on earth where no one will ever catch a cold and the freezing waters are so bitter there that a fish has been discovered to have developed its own anti-freeze. NOVA explores Antarctica—the coldest desert in the world.

Small Imperfection (A) 306
February 8, 1976
Producer(s): Robert Reid, BBC
Every year, some 5,000 babies are born in the US with spina bifida, a

congenital abnormality of the central nervous system. NOVA explores the mystery of what causes spina bifida and raises the issues of whether heroic measures should be taken to preserve the life of severely malformed babies.

Desert Place (A) 305
February 1, 1976
Producer(s): John Borden & Neil Goodwin, NOVA
NOVA explores the mysterious ecosystem of the desert: a snowstorm; a lashing summer monsoon; and the emergence—in a pool created only minutes before—of a pair of adult spadefoot toads who had been waiting beneath the sand for a year for this brief and fortuitous moment to procreate the next generation.

Planets (The) 304
January 25, 1976
Producer(s): Edward Goldwyn, BBC
The last fourteen years have been a revolution in our understanding of our place in the stars, the Solar System. Beginning in 1961 with a Russian spacecraft flying to Venus, quickening with the Apollo manned missions to the Moon, it came of age in the Spring of 1974, when there were six spacecrafts traveling simultaneously from the Earth to the planets. NOVA looks at the era of manned and unmanned exploration of the Solar System.

Meditation and the Mind 303
January 18, 1976
Producer(s): Graham Massey, BBC
What do singer Peggy Lee, New York Jets Quarterback Joe Namath and Congressman Richard Nolas have in common? They all practice a ritual called TM—Transcendental Meditation. NOVA examines the recent phenomenal success of the TM movement in America.

Joey 302
January 11, 1976
Producer(s): Brian Gibson, BBC
NOVA takes viewers into the world of Joey Deacon, 54 years old and a spastic since birth. Joey has lived most of his life in institutions, unable to communicate with anyone until he met Ernie Roberts. The docudrama recreates Joey's story, with remarkable performances by two spastic actors portraying him as a boy and as a young man. Joey and Ernie themselves appear in the final sequences.

Predictable Disaster 301
January 4, 1976
Producer(s): John Angier, NOVA
It is now possible to predict earthquakes. At least two successful predictions have already been made in the United States and the NOVA crew was present and filming while a third prediction was being formulated. NOVA looks at why earthquakes occur, how predictions are made, the threat they pose to cities at risk, and examines the advantages and disadvantages of making an earthquake a predictable disaster. [3rd season premiere]

Will The Fishing Have to Stop? 217
April 6, 1975
Producer(s): Ben Shedd, NOVA
Fish is an excellent source of protein; it could help ease the growing international food shortage. But in 1972 the total world fish catch dropped. NOVA explores the possible reasons for this decline.

Lost World of the Maya (The) 216
March 30, 1975
Producer(s): David Collison, BBC
For over a thousand years the Mayan civilization grew and flourished in the rain forests of Central America. Discovered and finally destroyed by the Spanish Conquistadors, it was lost until explorers brought it to light in the 19th century. Eric Thompson, an archeolo-

gist who has had a 45 year love affair with the Maya, takes NOVA on a pilgrimage through the Mayan world, visiting on the way, all the great ruined cities he has known for half a century.

Other Way (The) 215
March 16, 1975
Producer(s): John Mansfield, BBC
Since the Industrial Revolution, bigger has been better. NOVA profiles E.F. Schumacher, the author of Small is Beautiful, who thinks that enough is enough; that the time has come for technology to return to a human scale, where the ability to create is returned from the machine to people.

Plutonium Connection (The) 214
March 9, 1975
Producer(s): John Angier, NOVA
How likely is it that a terrorist group will steal plutonium intended for nuclear reactor fuel and put together a blackmail weapon of unprecedented power in the shape of a homemade atom bomb? That question is posed by Theodore Taylor, former A and H bomb designer at Los Alamos, in a recent book, The Curve of Binding Energy. NOVA investigates just how easy it would be to design a bomb using unclassified information.

Tuaregs (The) 213
February 16, 1975
Producer(s): Charlie Marin, Granada TV
High in the Hoggar Mountains, in the exact center of the Sahara desert, lives Sidi Mohammed and his family: children, grandchildren, cousins and a few former slave women. Their environment, one of the most ungenerous on earth, provides them with almost nothing. NOVA examines the changing lifestyle of Sidi Mohammed.

Lysenko Affair (The) 212
February 9, 1975
Producer(s): Peter Jones, BBC
NOVA explores T.D. Lynsenko's rise to power in the Soviet Union in the early 20th century, and how it affected plant genetic research in the USSR.

Take the World From Another Point of View 211
February 2, 1975
Producer(s): Francis Gladstone, NOVA
NOVA profiles two very different scientists: Richard Feynman, a theoretical physicist and Nobel prizewinner at the pinnacle of his career and Richard Lewontin, a biologist and highly regarded population geneticist from Harvard University.

Rise and Fall of DDT (The) 210
January 19, 1975
Producer(s): Alec Nisbett, BBC
Has the case against DDT been proven? A strange question, perhaps, to be asking one year after the US has banned the insecticide, but NOVA dares to ask. Tracing the history of DDT from its discovery through its banning in the States, NOVA asks whether America overreacted with its total ban of this once acclaimed "wonder" chemical.

What Time is Your Body? 209
January 12, 1975
Producer(s): Dominic Flessati, BBC
Have you ever sensed that your body reacts differently at different times of the day? NOVA examines the best and worst times for work, sex drives and your body's most reactive time of day for alcohol consumption.

War From the Air 208
January 5, 1975
Producer(s): Francis Gladstone, NOVA
NOVA explores how science and technology play a major role in the design of weapons of war and the development of strategies for their use.

Red Sea Coral 207
December 15, 1974
Producer(s): Alec Nisbett, BBC
NOVA joins a group of English biologists living literally on a platform in the middle of the Red Sea, who for several years have been studying the crown-of-thorns starfish, notorious for the devastation it has wrought on the coral reefs of Australia and the Pacific.

Men Who Painted Caves (The) 206
December 8, 1974
Producer(s): Christopher La Fontaine, BBC
Just why did Cro-Magnon man living in France's Dordogne Valley some 15,000 years ago take time out from the desperate business of survival to paint pictures in inaccessible corners of his cave dwellings? NOVA joins French and American archeologists as they piece together the lifestyle of these hunters of the last great Ice Age, and try to interpret the meaning of their cave art.

Inside the Golden Gate 205
December 1, 1974
Producer(s): John Angier, NOVA
NOVA joins a team of U.S. Geological Survey scientists on a mission to find out just how San Francisco Bay works: its physics, its chemistry and its biology.

Secrets of Sleep (The) 204
November 24, 1974
Producer(s): John Mansfield, BBC
Most of us spend one-third of our lives in a state of which we under-

stand remarkably little—some people sleep for only a few minutes a night, and function perfectly well, while others declare that eight hours isn't enough. NOVA explores traditional notions about how much sleep we need; looks at effects of the sleeping pill, and, perhaps the most baffling of all aspects of sleep—dreaming.

Hunting of the Quark (The) 203
November 17, 1974
Producer(s): David Paterson, BBC
Smashing matter into ever smaller pieces, particle accelerators attempt to find the fundamental building blocks of matter. NOVA looks at a team of physicists as they strive to find the elusive quark.

How Much Do You Smell? 202
November 10, 1974
Producer(s): Mick Rhodes, BBC
Many insects and some mammals use smell as a primary means of communication. NOVA explains how, for example, the entire economy of an ant's nest is organized by smell, and how some moths use smell for population control—an ability man is now beginning to understand.

Why Do Birds Sing? 201
November 3, 1974
Producer(s): Ben Shedd, NOVA
NOVA travels to forests and marshes to discover why birds sing and finds surprising parallels with the acquisition of speech in humans. [2nd season premiere]

Mystery of the Anasazi (The) 113
May 26, 1974
Producer(s): John Irving, NOVA
Who were the people that built the first cities—complete with apartment blocks—in North America? They were the Anasazi Indians, who lived in the Southwest for some eight or nine thousand years—and who then, in about 1300 AD, abruptly abandoned their cities

and apparently disappeared. NOVA traces the steps of this ancient sophisticated culture.

Fusion: The Energy of Promise 112
May 19, 1974
Producer(s): Stuart Harris & Mick Jackson, NOVA
Nuclear fusion offers the promises of an unlimited, clean source of energy. But achieving fusion has proved one of the most difficult and elusive goals of the physicist. NOVA tells the story of the twists and turns and the international competition along the road toward the achievement of fusion; and details the recent breakthroughs which seem at last to have brought it within reach.

Case of the Midwife Toad (The) 111
May 12, 1974
Producer(s): Bruce Norman, BBC
When Paul Kammerer committed suicide in 1926, it was taken by most of his fellow biologists as a tacit admission of guilt that he had faked his experiments purporting to show the inheritance of acquired characteristics. Arthur Koestler joins NOVA in an in-depth examination of Kammerer's infamous experiment.

First Signs of Washoe (The) 110
May 5, 1974
Producer(s): Simon Campbell Jones, NOVA
Washoe is a chimp more like a person: she talks with her hands. NOVA visits with Washoe and her teachers—Professor Allen Gardner and Dr. Trixie Gardner—to learn more about this unusual animal.

Are You Doing This for Me, Doctor? 109
April 28, 1974
Producer(s): Peter Jones, BBC
The advance of medicine depends inevitably on the testing of experimental procedures on human volunteers from either the healthy or the sick. Yet such procedures are often dangerous, and may not be of

direct benefit to the subject. NOVA examines how individuals' interests are safeguarded, and asks, under what circumstances experiments should be conducted on children.

Bird Brain: The Mystery of Bird Navigation 108
April 21, 1974
Producer(s): Tony Edwards, BBC
Birds migrate in search of perpetual summer, sometimes traveling as much as 20,000 miles every year. NOVA shows how using radar to track and identify migrating birds that travel at night can aid meteorologists in choosing routes that avoid bad weather and make the best use of prevailing winds.

Crab Nebula (The) 107
April 14, 1974
Producer(s): Alec Nisbett, BBC
In 1405 AD, the Chinese recorded the explosion of a star so bright that it lit the sky for three weeks, even during the day. It was the explosion of a dying star that was bigger than our sun. NOVA explores this mysterious explosion that led to the discovery of the Crab Nebula.

Strange Sleep 106
April 7, 1974
Producer(s): Francis Gladstone, NOVA
Medicine was transformed in the 19th century by the discovery of anesthesia. This NOVA docudrama depicts the pioneers of medicine and the impact that anesthesia has made on surgical procedures.

Last of the Cuiva 105
March 31, 1974
Producer(s): Brian Moser, Granada TV
How does a primitive nomadic tribe of the Amazon basin cope with the encroachment of Western settlers? NOVA looks at both sides of the story, revealing the misunderstandings between the two cultures.

Search for Life (The) 104
March 24, 1974
Producer(s): John Angier, NOVA
Does life exist outside this planet? The Viking Lander will set down on Mars in July 1976 to try to find out just that. NOVA explores how life started on Earth and examines the Viking Lander being built in its germ-free room before starting its long journey.

Whales, Dolphins, and Men 103
March 17, 1974
Producer(s): Simon Campbell Jones, BBC
NOVA explores the impact of whaling and the goods it produces for the industry.

Where Did Colorado Go? 102
March 10, 1974
Producer(s): Simon Campbell Jones, NOVA
NOVA explores the mighty Colorado River which today has become the life-blood of the Southwest, providing water and electricity to the farms and cities of California, Nevada, and Arizona. The program examines the political expediency and technological over-optimism that has led to some major miscalculations of the river's capacity.

Making of a Natural History Film (The) 101
March 3, 1974
Producer(s): Mick Rhodes, BBC
NOVA premieres on public television with a behind-the-scenes look at the making of a nature film. Oxford Scientific Films Unit shows how it tackles such problems as filming a wood-wasp laying its egg inside trees, the hatching of a chick and the courtship rituals of the stickleback. [1st season premiere]